Macromolecular Symposia 209

Organometallic and Coordination Clusters and Polymers: Syntheses and Applications

London, Ontario, Canada
May 29–June 1, 2004

Symposium Editors:
A.S. Abd-El-Aziz, P.D. Harvey, Winnipeg, Canada

pp. 1–251 · March 2004
ISBN 3-527-31042-8

Macromolecular Symposia publishes lectures given at international symposia and is issued irregularly, with normally 14 volumes published per year. For each symposium volume, an Editor is appointed. The articles are peer-reviewed. The journal is produced by photo-offset lithography directly from the authors' typescripts.
Further information for authors can be found at http://www.ms-journal.de.
Suggestions or proposals for conferences or symposia to be covered in this series should also be sent to the Editorial office (E-mail: macro-symp@wiley-vch.de).

Editor: Ingrid Meisel
Senior Associate Editor: Stefan Spiegel
Associate Editor: Alexandra Carrick
Assistant Editors: Sandra Kalveram,
                        Mara Staffilani

Executive Advisory Board:
M. Antonietti (Golm), M. Ballauff (Bayreuth), S. Kobayashi (Kyoto), K. Kremer (Mainz), T. P. Lodge (Minneapolis), H. E. H. Meijer (Eindhoven), R. Mülhaupt (Freiburg), A. D. Schlüter (Berlin), J. B. P. Soares (Waterloo), H. W. Spiess (Mainz), G. Wegner (Mainz)

Macromolecular Symposia:
Annual subscription rates 2004
Macromolecular Full Package: including Macromolecular Chemistry & Physics (18 issues), Macromolecular Rapid Communications (24), Macromolecular Bioscience (12), Macromolecular Theory & Simulations (9), Macromolecular Materials and Engineering (12), Macromolecular Symposia (14):

Europe	Euro	6.424 / 7.067
Switzerland	Sfr	11.534 / 12.688
All other areas	US$	7.948 / 8.743

print only **or** electronic only / print **and** electronic

Postage and handling charges included. All Wiley-VCH prices are exclusive of VAT. Prices are subject to change.

Single issues and back copies are available. Please ask for details at: service@wiley-vch.de.

Orders may be placed through your bookseller or directly at the publishers:
WILEY-VCH Verlag GmbH & Co. KGaA, P. O. Box 10 11 61, 69451 Weinheim, Germany, Tel. +49 (0) 62 01/6 06-400, Fax +49 (0) 62 01/60 61 84. E-mail: service@wiley-vch.de

For USA and Canada: Macromolecular Symposia (ISSN 1022-1360) is published with 14 volumes per year by WILEY-VCH Verlag GmbH Co. KGaA, Boschstr. 12, 69451 Weinheim, Germany. Air freight and mailing in the USA by Publications Expediting Inc., 200 Meacham Ave., Elmont, NY 11003, USA. Application to mail at Periodicals Postage rate is pending at Jamaica, NY 11431, USA. POSTMASTER please send address changes to: Macromolecular Symposia, c/o Wiley-VCH, III River Street, Hoboken, NJ 07030, USA.

Macromolecular Symposia

Articles published on the web will appear several weeks before the print edition. They are available through:

www.ms-journal.de

www.interscience.wiley.com

Organometallic and Coordination Clusters and Polymers: Syntheses and Applications Symposium

London, Ontario (Canada), 2004

Preface
A. S. Abd-El-Aziz, P. D. Harvey

Author Index

Preface

Forty nine years ago, the first ferrocene-based polymer was reported by Arimoto and Haven at the Dupont Co. and since that time the field of organometallic polymers has grown tremendously. This class of polymers exhibits remarkable electrochemical, morphological, pharmacological, optical, magnetic and thermal properties. While coordination polymers are one of the earliest developed examples of metal-containing polymers, they continue to attract great attention due to their widespread applications.

This book is a result of the Symposium titled "Organometallic and Coordination Clusters and Polymers: Syntheses and Applications" held in London, Ontario, Canada at the 87th Canadian Chemistry Conference and Exhibition, May 28th to June 1st, 2004. It contains articles that provide an overview of this field and covers recent developments in the field of organometallic and coordination clusters and polymers. These articles were written by some of the leading researchers in this field from Australia, Japan, the United Kingdom, the United States and Canada, and deal with hot topics such as ring-opening-polymerization, conducting and photo-conducting organometallic polymers, heterogenized homogeneous catalysis, organometallic dendrimers, self-assemblies of organometallic/coordination complexes including metallo-DNA, unusual photophysical properties and electroluminescence of organometallic and coordination dendrimers, oligomers and polymers, molecular dimension determination of small oligomers, oligoclusters en route towards polymers of clusters, azo dye functionalized polymers, photochemical degradation, and recent applications for lithography. The aim of this symposium is to generate dialogue between these researchers and to allow them to present some of their most recent research.

This symposium was supported by the Inorganic Chemistry Division of the Canadian Society for Chemistry. We would like to extend our deepest appreciation to all of the contributors to this volume for their willingness to prepare their manuscripts in advance and for their participation in this annual event.

A. S. Abd-El-Aziz
P. D. Harvey

Towards Oligoclusters – Chemistry and Properties

Mark G. Humphrey

Department of Chemistry, Australian National University, Canberra, ACT 0200, Australia

E-mail: mark.humphrey@anu.edu.au

Summary: The recent progress of the author's research group in the syntheses, electronic properties and optical properties of monomeric and oligomeric organotransition metal cluster-containing systems is reviewed.

Keywords: cluster; electrochemistry; mixed-metal; oligomer; optical limiting

Introduction

A determined worldwide push has clarified several aspects of organotransition metal cluster chemistry - particularly well-established are synthetic methods to small homometallic clusters (those with three- or four-metal-atom cores), and aspects of their reaction chemistry.[1] Considerably less well-established, but of topical interest, are protocols to assemble clusters into oligocluster arrays and the physical properties of clusters (e.g. electronic, optical, magnetic, etc.). Significant interest has been shown recently in organotransition metal-containing polymers[2] and in metallodendrimers and metallostars,[3] metal-containing branched molecules whose molecular architectures offer prospects of a range of new applications. Thus far, however, few cluster-containing examples are extant. Some recent work has emphasized network structures based on the $Re_6(\mu_3-E)_8L_6$ core: see, for example, Ref. [4]. Cluster-containing oligomers are also relatively rare, despite the fact that such oligocluster arrays promise processable functional materials (note that many studies have examined the coordination of metal clusters to prefunctionalized polymers, mainly through P-linkages,[5-15] but also *via* O-[16] C-,[17-19] and N-interactions;[20] however, polymer supports of this type frequently result in polycoordination,[9] affording a mixture of mono- and poly-substituted clusters, together with non-coordinated polymer-bound ligands attached to the backbone). Only a few examples of oligomers with clusters in the oligomer backbone are extant.[21-24] Most cluster-containing polymers contain cluster units linked by P- or N-ligands or isocyanides, all of which can potentially be displaced leading to polymer breakdown, and all examples except $\{Ru_6(\mu_6\text{-}C)(CO)_{15}(Ph_2PC_2PPh_2)\}_n$ are insoluble - new routes to soluble cluster-containing polymers are clearly required. Cluster-containing dendrimers and "star" molecules have also been the subject of very few reports,[25-27] although their metal-rich composition and electro-active nature may be useful in materials applications. In contrast, clusters linked through π-delocalizable backbones have come under considerable scrutiny, but the vast majority of such studies involve the linking of identical homometallic cores.[28-40] Introduction of a heterometal into the cluster core can further modify the electronic environment,[41] but very few examples of heterometallic clusters linked by unsaturated bridges exist.[42-44] Reported herein are our syntheses of a cluster-centered "star" molecule, cluster-containing oligourethanes, and mixed-metal oligo-cluster species in which the cluster units are linked by potentially π-delocalizable bridges.

Optical limiters are materials that display decreasing transmittance as a function of incident light intensity (i.e. their optical transmission is power dependent). Efficient optical limiters are required for a range of applications in optical device protection - for example, such materials may be used for the protection of eyes and sensors from intense light pulses, as well as in laser mode locking and optical pulse shaping. Clusters (with fullerenes and phthalocyanines) have been identified as one of three promising classes of molecular optical limiters; all three can be viewed as reduced dimensionality materials with confined but potentially highly delocalizable electrons.[45] Optical limiting can arise from several phenomena (e.g. two-photon absorption (TPA), reverse saturable absorption (RSA)). The difference between the RSA process and TPA is that TPA is virtually instantaneous whereas processes involving an intermediate absorbing state exhibit certain kinetic behavior, which is dependent on the lifetimes of the states involved. Time-resolved investigations of the changes of absorptive properties are necessary to evaluate the mechanism of power limiting in a given system. The challenge at present is to develop an understanding of the underlying physical properties which give rise to the desired responses, and to design robust readily processable materials exhibiting optical limiting properties. Also summarized herein are our studies of the optical limiting behavior of systematically-varied mixed-metal clusters and a cluster-containing oligourethane.

Results and Discussion

A cluster-centered "star". Dendrimers with clusters at the periphery are examples of nanoscale molecular architectures of considerable current interest,[46-48] but branched molecules with clusters at the core are little-explored. The well-established facile tris-substitution of $Ru_3(CO)_{12}$ by *P*-donor ligands suggests that it could function as a core unit and branching point in dendrimer construction and, indeed, that with the use of appropriate bifunctional ligands one can construct cluster-centered "star" molecules. With this in mind, we synthesized new phosphines bearing functional alkynyl units to facilitate dendrimer and star formation. The new phosphines $Ph_2PC_6H_4-4-C\equiv CR$ [R = $SiMe_3$, H] were used to prepare $Ru(C\equiv CC_6H_4-4-PPh_2)(PPh_3)_2(\eta-C_5H_5)$ (**1**) (Scheme 1) and $Ru_3(CO)_9(Ph_2PC_6H_4-4-C\equiv CSiMe_3)_3$ (**2**) (Scheme 2), respectively, the former with a pendent phosphine. Reaction of **2** with carbonate or fluoride gave $Ru_3(CO)_9(Ph_2PC_6H_4-4-C\equiv CH)_3$ (**3**) with pendent terminal alkyne groups (Scheme 2). Reaction of **3** with [$Ru(NCMe)(PPh_3)_2(\eta-C_5H_5)$]$PF_6$ or reaction of $Ru_3(CO)_{12}$ with **1** afforded $Ru_3(CO)_9\{(Ph_2PC_6H_4-4-C\equiv C)Ru(PPh_3)_2(\eta-C_5H_5)\}_3$, a cluster-centered star (Scheme 3).[49]

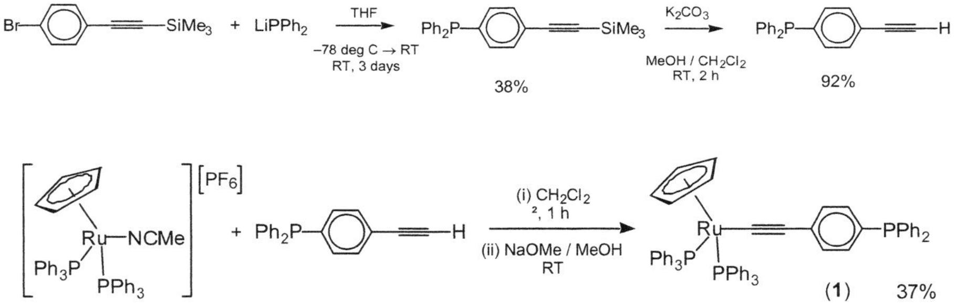

Scheme 1. Synthesis of $Ru(C\equiv CC_6H_4-4-PPh_2)(PPh_3)_2(\eta-C_5H_5)$ (**1**).

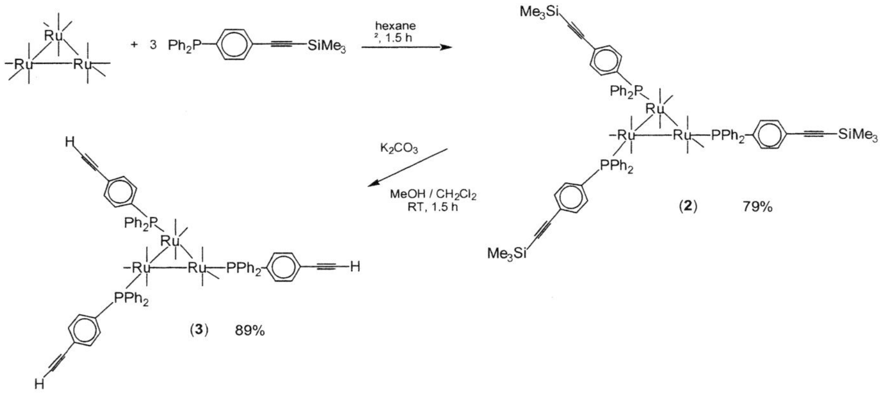

4

hexane
², 1.5 h

3 Ph_2P—⬡—$SiMe_3$

Me_3Si

(2) 79%

K_2CO_3

MeOH / CH_2Cl_2
RT, 1.5 h

(3) 89%

Scheme 2. Synthesis of $Ru_3(CO)_9(Ph_2PC_6H_4\text{-}4\text{-}C\equiv CSiMe_3)_3$ (2) and $Ru_3(CO)_9(Ph_2PC_6H_4\text{-}4\text{-}C\equiv CH)_3$ (3).

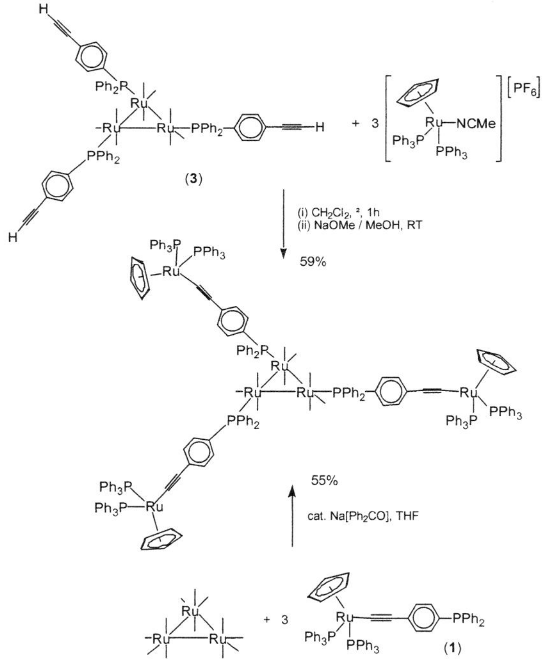

(3)

$+ 3$ $\left[\begin{array}{c} Ru\text{—NCMe} \\ Ph_3P \quad PPh_3 \end{array}\right] [PF_6]$

(i) CH_2Cl_2, ², 1h
(ii) NaOMe / MeOH, RT

59%

55%

cat. Na[Ph_2CO], THF

$+ 3$ (1)

Scheme 3. Syntheses of $Ru_3(CO)_9\{(Ph_2PC_6H_4\text{-}4\text{-}C\equiv C)Ru(PPh_3)_2(\eta\text{-}C_5H_5)\}_3$.

Cluster-containing oligourethanes. As was mentioned above, one particular problem with previously-reported cluster-containing polymers is the ease of polymer breakdown. Replacing the *P*- or *N*-ligands with substituted cyclopentadienyl ligands, which are significantly more strongly bound to clusters, should remedy this problem. Polyamides, polyurethanes and polyureas containing photodegradable $Mo_2(CO)_6(\eta\text{-}C_5H_4R)_2$ units along the polymer backbone have been reported previously.[50-56] We replaced the photo-active $Mo_2(CO)_6(\eta\text{-}C_5H_4R)_2$ groups in these polymers with photo-stable cluster units, affording the first cluster-containing oligomers in which the clusters are in the structurally-uniform environment of the oligomer backbone and attached via robust cyclopentadienyl groups.[57] Bis(hydroxyalkylcyclopentadienyl)-containing mixed molybdenum-iridium clusters $Mo_2Ir_2(CO)_{10}\{\eta\text{-}C_5H_4(CH_2)_xOH\}_2$ [x = 2 (**4**), 10] were reacted with alkyl or aryl 1,ω-diisocyanates OCNRNCO [R = $(CH_2)_y$ (y = 4, 6, 12), *trans*-1,4-cyclohexyl, or 4-$C_6H_4CH_2$-4-C_6H_4] to form oligourethanes with transition metal clusters in the oligomer backbone (Scheme 4; 46-89% yields).

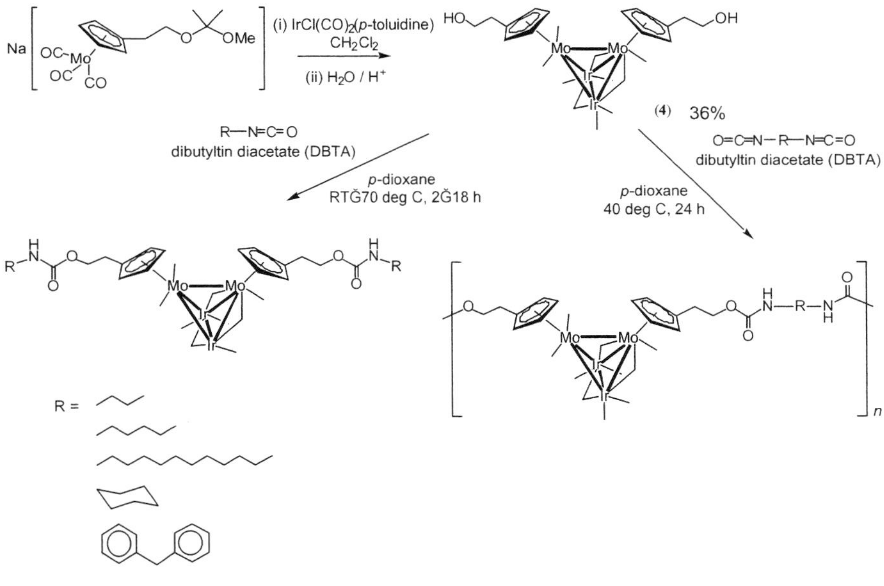

Scheme 4. Preparation of oligourethanes containing clusters in the backbone and model compounds.

Characterization of the cluster-containing oligourethanes was aided by spectral comparison with model cluster-diurethanes $Mo_2Ir_2(CO)_{10}\{\eta\text{-}C_5H_4(CH_2)_2OC(O)NHRH\}_2$ [R = $(CH_2)_y$ [y = 4, 6, 12], *trans*-1,4-cyclohexyl; 37-72% yields] prepared from reaction between the cluster diol **4** and alkyl isocyanates HRNCO (Scheme 4). The extent of polymerization was assessed by gel permeation chromatography, with little dependence on diisocyanate precursor linker R, but strong dependence on alkylcyclopentadienyl linker length $(CH_2)_x$, suggesting that the steric influence of the bulky dimolybdenum-diiridium cluster core and co-ligands is the most important factor governing extent of polymerization.

π-Delocalizable bridge-linked clusters. Following these studies of cluster-containing oligourethanes, in which the cluster units are linked via saturated bridges, our attention was drawn to the possibility of linking such clusters by unsaturated bridging groups, initial studies focussing on di- and tri-cluster units linked by a variety of potentially π-delocalizable bridges (phenylene, phenylenevinylene, phenyleneethynylene, thienyl, selenienyl); ligation of the π-bridge to the cluster core was by a $\mu_4\text{-}\eta^2$-bound alkyne, a particularly robust interaction. Our first attempt employed a cyclopentadienyl-containing cluster and a di-terminal alkyne (Scheme 5), but the limited solubility of the product complicated purification, so subsequent studies utilized methylcyclopentadienyl-containing clusters and n-hexyl-containing internal alkynes, dramatically enhancing solubility.[58]

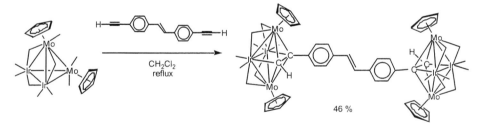

Scheme 5. Preparation of a *trans*-stilbene linked dicluster compound.

Alkynes react with these tetrahedral dimolybdenum-diiridium or ditungsten-diiridium clusters by formal insertion of the C_2 unit into the Mo-Mo or W-W bond to afford robust clusters with pseudooctahedral core geometries. Because we were interested in examining the modification of properties upon linking clusters by π-delocalizable bridges, we prepared several mono-cluster model compounds to benchmark an isolated cluster unit, together with mono-cluster compounds with functionalized alkynes suitable for coupling to form di-cluster compounds (Scheme 6).[58]

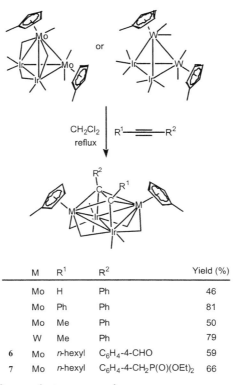

	M	R^1	R^2	Yield (%)
	Mo	H	Ph	46
	Mo	Ph	Ph	81
	Mo	Me	Ph	50
	W	Me	Ph	79
6	Mo	n-hexyl	C$_6$H$_4$-4-CHO	59
7	Mo	n-hexyl	C$_6$H$_4$-4-CH$_2$P(O)(OEt)$_2$	66

Scheme 6. Syntheses of mono-cluster compounds.

Similar reactions between tetrahedral cluster precursors and vinylene-linked di- or tri-ynes afforded related di- or tri-cluster compounds in which two or three cluster units are linked by unsaturated bridges (Schemes 7-9).[58]

8

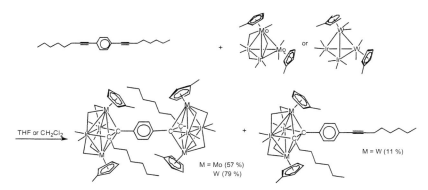

Scheme 7. Syntheses of phenylene-linked di-cluster compounds.

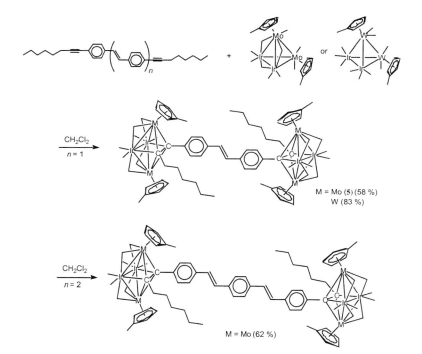

Scheme 8. Syntheses of phenylenevinylene-linked di-cluster compounds.

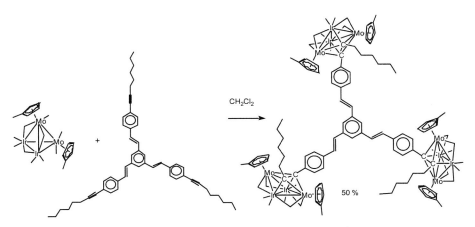

Scheme 9. Synthesis of a phenylenevinylene-linked tri-cluster compound.

Compound **5** was also prepared by coupling **6** and **7** under Emmons-Horner conditions, but in a lower yield (Scheme 10).[58]

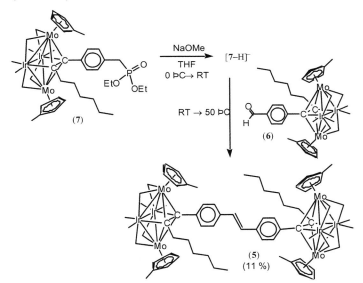

Scheme 10. Synthesis of a di-cluster compound by Emmons-Horner coupling.

Reactions between the tetrahedral mixed-metal clusters and ethynylene-linked di- or triynes afforded related mono-, di- or tri-cluster compounds (Schemes 11 and 12). Attempts to add a third cluster to the 1,3,5-tris(1-octynyl)benzene were unsuccessful, presumably due to steric reasons.[59]

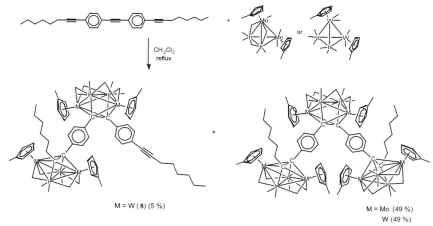

Scheme 11. Synthesis of phenyleneethynylene-linked di- and tri-cluster compounds.

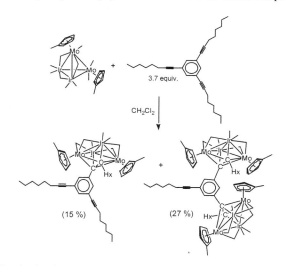

Scheme 12. Synthesis of 1,3,5-tris(1-octynyl)benzene mono- and di-cluster compounds.

Compound **8** corresponds to the 1,2-discluster adduct of the linear triyne Me(CH$_2$)$_5$C≡C-4-C$_6$H$_4$C≡C-4-C$_6$H$_4$C≡C(CH$_2$)$_5$Me. No 1,3-discluster isomer was isolated from direct reaction, but the related molybdenum-containing 1,3-discluster isomer was prepared by exploiting organic reaction chemistry on pre-coordinated functionalized alkyne ligands (Schemes 13 and 14).

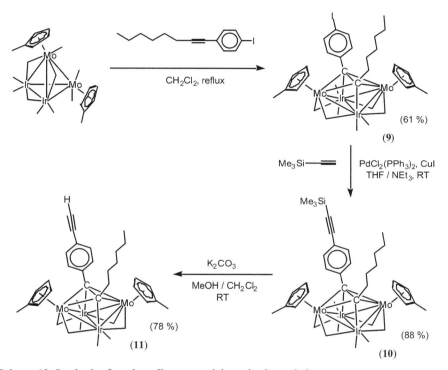

Scheme 13. Synthesis of pendent alkyne-containing mixed-metal clusters.

Sonogashira coupling of **9** with Me$_3$SiC≡CH and subsequent desilylation afforded Mo$_2$Ir$_2$\{μ$_4$-η^2-Me(CH$_2$)$_5$C$_2$-4-C$_6$H$_4$C≡CR\}(CO)$_8$(η^5-C$_5$H$_4$Me)$_2$ [R = SiMe$_3$ (**10**), H (**11**)]. Sonogashira coupling of **9** and **11** gave the 1,3-isomer [Mo$_2$Ir$_2$(CO)$_8$(η^5-C$_5$H$_4$Me)$_2$]$_2$\{μ$_8$-η^4-Me(CH$_2$)$_5$C$_2$-4-C$_6$H$_4$C≡CC$_6$H$_4$-4-C$_2$(CH$_2$)$_5$Me\}, as well as the homo-coupling product.[59]

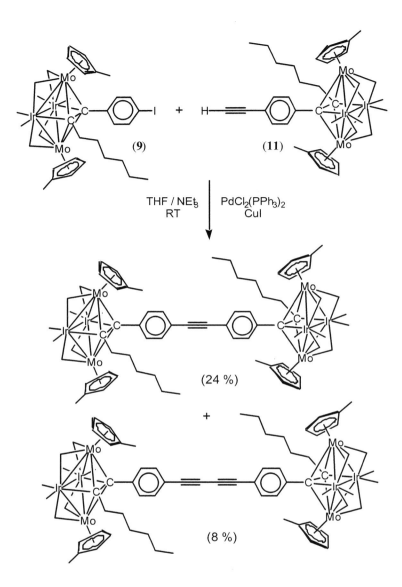

(9) + **(11)**

THF / NEt₃   PdCl₂(PPh₃)₂
RT   CuI

(24 %)

+

(8 %)

Scheme 14. Synthesis of a 1,3-dicluster isomer and homocoupled product.

Compounds with heterocycle-containing bridges are accessible by similar procedures. The tetrahedral mixed-metal clusters reacted with heterocyclic di- or triynes to afford the analogous mono-, di- or tri-cluster compounds to the phenyl-containing examples above (Schemes 15-17).[60]

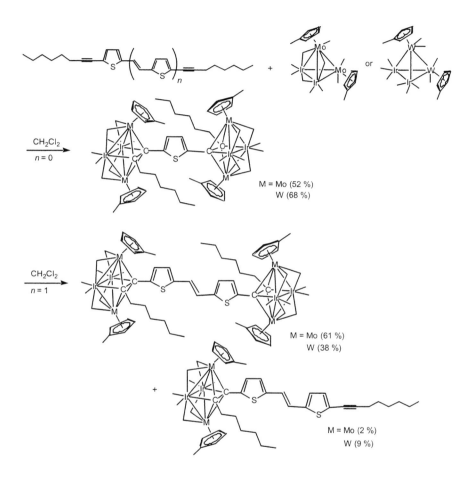

Scheme 15. Syntheses of thienyl-linked di-cluster compounds.

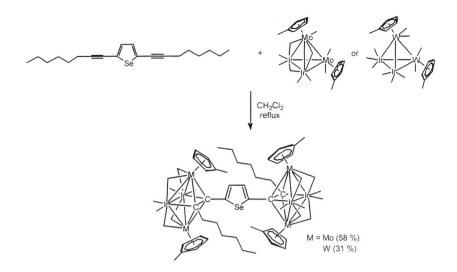

Scheme 16. Synthesis of selenienyl-linked di-cluster compounds.

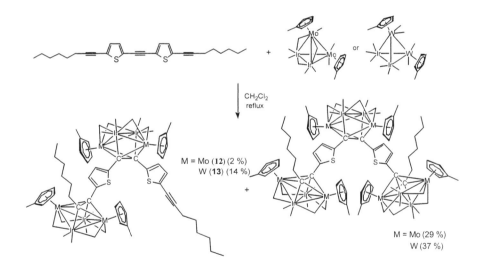

Scheme 17. Syntheses of thienylethynylene-linked di- and tri-cluster compounds.

Compounds **12** and **13** correspond to the 1,2-dicluster adducts of the linear triyne Me(CH$_2$)$_5$C≡C-5-C$_4$H$_2$S-2-C≡C-2-C$_4$H$_2$S-5-C≡C(CH$_2$)$_5$Me. As with the analogous phenyl-based chemistry, no 1,3-dicluster isomer was isolated from direct reaction, but the molybdenum-containing 1,3-dicluster isomer was prepared by exploiting organic reaction chemistry on pre-coordinated functionalized alkyne ligands (Schemes 18 and 19).[60]

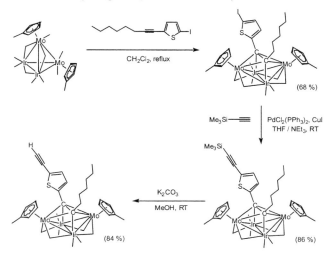

Scheme 18. Synthesis of pendent alkyne-containing mixed-metal clusters.

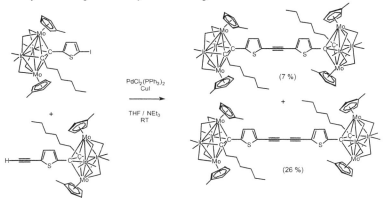

Scheme 19. Synthesis of a 1,3-dicluster isomer and homocoupled product.

A di-cluster compound with a saturated bridge was prepared in a similar manner (Scheme 20).[58]

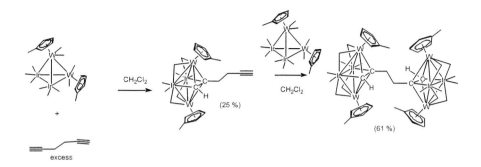

Scheme 20. Synthesis of a di-cluster compound with a saturated bridging group.

The properties of these systematically-varied oligo-cluster compounds are currently being studied. Cyclic voltammetric scans for many examples have been collected. All compounds show a reversible/partially-reversible oxidation, followed by an irreversible oxidation process; potentials for the former increase on replacement of tungsten by molybdenum and alkyne substituent variation Me < H < Ph. UV-vis-NIR spectroelectrochemical studies of the first oxidation process show similar spectral progressions for mono- and di-cluster compounds. The reductive cyclic voltammetric scans for compounds in which clusters are linked by long unsaturated bridges show one irreversible reduction process, whereas compounds in which n cluster cores are linked by a phenylene unit show n irreversible reduction processes. Density functional calculations indicate that oxidation and reduction both proceed with retention of the pseudooctahedral core geometry, but that loss of a carbonyl ligand concomitant with two-electron reduction is energetically accessible, suggesting that this accounts for the irreversibility of the reduction step.[58-60]

Optical limiting properties of clusters. Metal carbonyl clusters tend to be obtained as crystalline materials rather than in a film-processable form required for putative optical applications. However, access to oligomers and polymers with clusters in the backbone should afford processable cluster-containing materials where the clusters are in the uniform chemical environments required to standardize molecular optical properties. We have commenced a study of the optical limiting behavior of our mixed-metal clusters; this has involved development of preliminary structure-activity trends, and examination of the temporal nature of the optical limiting response in this class of optical limiter.[61]

Linear optical absorption spectra were obtained. Progression from tetrahedral molybdenum-triiridium cluster to dimolybdenum-diiridium cluster and replacing molybdenum by tungsten both shift the optical absorption maxima to lower energy. All clusters contain broad low intensity absorptions through the visible region, suggestive of potential as broad-band optical limiters. Incorporation of the cluster into the polymer does not result in modification of the optical spectrum, consistent with electronically "insulated" clusters, as expected, permitting optical properties of the cluster to be exploited while processability is enhanced.

The optical limiting properties were assessed by open-aperture Z-scan (ns pulses, 523 nm) and time-resolved pump-probe studies (ps pulses, 527 nm). A typical data set and ideal fitted curve from the former is shown in Figure 1.

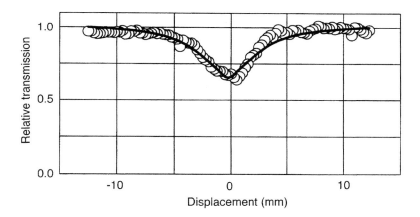

Figure 1. Z-scan data set and ideal fitted curve for $Mo_2Ir_2(CO)_{10}(\eta\text{-}C_5Me_5)_2$.

We have utilized the effective nonlinear absorption coefficient, β_2, which is extracted from the curve and extrapolated to pure substance, as an indicator of optical limiting potential. Results from tetrahedral mixed group 6 - iridium clusters are displayed in Table 1.

Table 1. Optical limiting results from open aperture Z-scan experiments

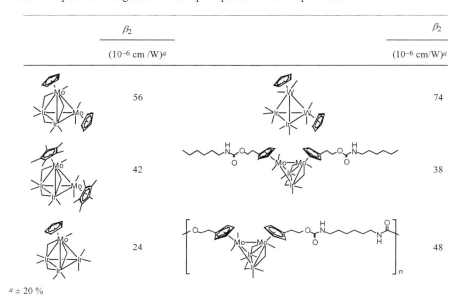

	β_2		β_2
	$(10^{-6}\ cm\ /W)^a$		$(10^{-6}\ cm/W)^a$
	56		74
	42		38
	24		48

$^a \pm 20\ \%$

Several trends in β_2 are revealed by this study (while being mindful of the error margins). The molybdenum-triiridium cluster exhibited a lower nonlinear absorption coefficient than its dimolybdenum-diiridium cluster analogue. Alkylation of the cyclopentadienyl ring also decreased this coefficient, although less significantly so. Replacing tungsten by molybdenum resulted in a decrease in β_2. The incorporation of the clusters into polymers increases processability, but does not appear to diminish nonlinear absorption, consistent with the optical limiting response originating from the ligated metal core, and with no inter-cluster electronic interaction to influence optical limiting.

Employing nanosecond laser pulses to measure nonlinear absorption can result in many time-dependent photophysical processes becoming integrated into the response. We therefore employed a time-resolved pump-probe experiment, from which one can derive excited state lifetimes (τ) and absorption cross sections ('ease of absorption' between two states, σ). The results from a 527 nm pump-probe experiment (pulses ca 50 ps, pump fluence at the sample *ca*

1 J cm^{-2}) on a solution of $Mo_2Ir_2(\mu_4\text{-}\eta^2\text{-}MeC_2Ph)(CO)_8(\eta\text{-}C_5H_4Me)_2$ are shown in Figure 2.

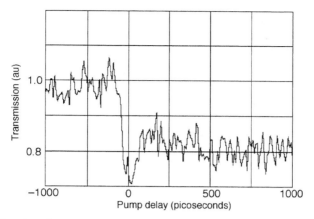

Figure 2.   Results from a 527 nm ps pump-probe experiment on $Mo_2Ir_2(\mu_4\text{-}\eta^2\text{-}MeC_2Ph)(CO)_8(\eta\text{-}C_5H_4Me)_2$.

The negative peak at 0 ps is likely to arise from an instantaneous TPA phenomenon when the probe beam arrives simultaneously with the pump beam. When the probe beam arrives after the pump beam, a long-lived tail is observed (transmission ≈ 0.8), likely to result from excited state absorption and the formation of metastable states; the lower energy excited states (metastable states) are sufficiently populated by the intense pump beam that absorption of photons from the probe beam (to higher energy excited states) becomes possible, resulting in a lower transmission. Figure 2 shows that the metastable states, required for the excited state absorption, are long-lived (> 1000 ps). The alkyne-adduct $Mo_2Ir_2(\mu_4\text{-}\eta^2\text{-}MeC_2Ph)(CO)_8(\eta\text{-}C_5H_4Me)_2$ can thus be classed as an optical power limiter exhibiting both TPA and RSA.

Conclusions

The studies summarized herein have demonstrated the facile synthesis of a variety of di-, tri- and oligo-cluster compounds of linear and branched geometries and with saturated and unsaturated bridging groups. Structure-optical limiting behavior relationships for the tetrahedral core mixed-metal clusters were developed. A time-resolved investigation of the alkyne-adduct $Mo_2Ir_2(\mu_4\text{-}\eta^2\text{-}$

$MeC_2Ph)(CO)_8(\eta\text{-}C_5H_4Me)_2$ using ps pulses at 527 nm revealed optical power limiting behavior resulting from electronic processes [a fast nonlinear absorption process followed by reverse saturable absorption involving long-lived (>1000 ps) metastable excited states].

[1] P. Braunstein, L. A. Oro, P. R. Raithby, *"Metal Clusters in Chemistry"*, Wiley-VCH, Weinheim, Germany, 1999.
[2] P. Nguyen, P. Gómez-Elipe, I. Manners, *Chem. Rev.* **1999**, *99*, 1515.
[3] M. A. Hearshaw, J. R. Moss, *J. Chem. Soc., Chem. Commun.* **1999**, 1.
[4] M. V. Bennett, L. G. Beauvais, M. P. Shores, J. R. Long, *J. Am. Chem. Soc.*, **2001**, *123*, 8022.
[5] K. Iwatate, S. R. Dasgupta, R. L. Schneider, G. C. Smith, K. L. Watters, *Inorg. Chim. Acta* **1975**, *15*, 191.
[6] M. S. Jarrell, B. C. Gates, *J. Catal.* **1978**, *54*, 81.
[7] M. S. Jarrell, B. C. Gates, E. D. Nicholson, *J. Am. Chem. Soc.* **1978**, *100*, 5727.
[8] J. J. Rafalko, J. Lieto, B. C. Gates, G. L. Schrader Jr, *J. Chem. Soc., Chem. Commun.* **1978**, 540.
[9] J. Lieto, J. J. Rafalko, B. C. Gates, *J. Catal.* **1980**, *62*, 149.
[10] Z. Otero-Schipper, J. Lieto, B. C. Gates, *J. Catal.* **1980**, *63*, 175.
[11] D. F. Foster, J. Harrison, B. S. Nicholls, A. K. Smith, *J. Organomet. Chem.* **1983**, *248*, C29.
[12] M. Wolf, J. Lieto, B. A. Matrana, D. B. Arnold, B. C. Gates, H. Knözinger, *J. Catal.* **1984**, *89*, 100.
[13] J. Lieto, M. Wolf, B. A. Matrana, M. Prochazka, B. Tesche, H. Knözinger, B. C. Gates, *J. Phys. Chem.* **1985**, *89*, 991.
[14] D. F. Foster, J. Harrison, B. S. Nicholls, A. K. Smith, *J. Organomet. Chem.* **1985**, *295*, 99.
[15] E. Paetzold, H. Pracejus, G. Oehme, *J. Mol. Catal.* **1987**, *42*, 301.
[16] J. Lieto, M. Prochazka, D. B. Arnold, B. C. Gates, *J. Molec. Catal.* **1985**, *31*, 89.
[17] S. Bhaduri, H. Khwaja, K. R. Sharma, *Indian J. Chem.* **1982**, *21A*, 155.
[18] H.P. Withers Jr, D. Seyferth, *Inorg. Chem.* **1983**, *22*, 2931.
[19] B. F. G. Johnson, A. J. Blake, A. J. Brown, S. Parsons, P. Taylor, *J. Chem. Soc., Chem. Commun.* **1995**, 2117.
[20] S. Bhaduri, H. Khwaja, B. A. Narayanan, *J. Chem. Soc., Dalton Trans.* **1984**, 2327.
[21] S. Aime, R. Gobetto, G. Jannon, D. Osella, *J. Organomet. Chem.* **1986**, *309*, C51.
[22] R. Fuchs, P. Klüfers, *J. Organomet. Chem.* **1992**, *424*, 353.
[23] A. M. Bradford, E. Kristof, M. Rashidi, D.-S. Yang, N. C. Payne, R. J. Puddephatt, *Inorg. Chem.* **1994**, *33*, 2355.
[24] B. F. G. Johnson, K. M. Sanderson, D. S. Shephard, D. Ozkaya, W. Zhou, H. Ahmed, M. D. R. Thomas, L. Gladden, M. Mantle, *Chem. Commun.* **2000**, 1317.
[25] C. B. Gorman, B. L. Parkhurst, W. Y. Su, K.-Y. Chen, *J. Am. Chem. Soc.*, **1997**, *119*, 1141.
[26] E. C. Constable, O. Eich, C. E. Housecroft, L. A. Johnston, *Chem. Commun.* **1998**, 2661.
[27] E. C. Constable, C. E. Housecroft, B. Krattinger, M. Neuburger, M. Zehnder, *Organometallics* **1999**, *18*, 2565.
[28] G. H. Worth, B. H. Robinson and J. Simpson, *Organometallics* **1992**, *11*, 3863.
[29] D. Osella, O. Gambino, C. Nevi, M. Ravera, D. Bertolino, *Inorg. Chim. Acta* **1993**, *206*, 155.
[30] S. M. Elder, B. H. Robinson, J. Simpson, *J. Organomet. Chem.* **1990**, *398*, 165.
[31] J. R. Fritch, K. P. C. Vollhardt, *Angew. Chem. Int. Ed. Engl.* **1980**, *19*, 559.
[32] G. H. Worth, B. H. Robinson, J. Simpson, *Organometallics* **1992**, *11*, 501.
[33] R. J. Dellaca, B. R. Penfold, B. H. Robinson, W. T. Robinson, J. L. Spencer, *Inorg. Chem.* **1970**, *9*, 2204.
[34] R. J. Dellaca and B. R. Penfold, *Inorg. Chem.* **1971**, *10*, 1269.
[35] M. I. Bruce, M. L. Williams, J. M. Patrick, A. H. White, *J. Chem. Soc., Dalton Trans.* **1985**, 1229.
[36] C. J. Adams, M. I. Bruce, E. Horn, B. W. Skelton, E. R. T. Tiekink, A. H. White, *J. Chem. Soc., Dalton Trans.* **1993**, 3299.
[37] J.-C. Daran, E. Cabrera, M. I. Bruce, M. L. Williams, *J. Organomet. Chem.* **1987**, *319*, 239.
[38] W. Y. Wong, S. H. Cheung, S. M. Lee, S. Y. Leung, *J. Organomet. Chem.* **2000**, *596*, 36.
[39] D. Osella, J. Hanzlík, *Inorg. Chim. Acta* **1993**, *213*, 311.
[40] D. Osella, O. Gambino, C. Nervi, M. Ravera, M. V. Russo, G. Infante, *Gazz. Chim. Ital.* **1993**, *123*, 579.

[41] S. M. Waterman, N. T. Lucas, M. G. Humphrey, In *"Adv. Organomet. Chem."*, A. Hill, R. West, Eds., Academic Press, London 2000, Vol. 46, p 47.

[42] M. I. Bruce, J.-F. Halet, S. Kahlal, P. J. Low, B. W. Skelton, A. H. White, *J. Organomet. Chem.* **1999**, *578*, 155.

[43] M. P. Jensen, D. A. Phillips, M. Sabat, D. F. Shriver, *Organometallics* **1992**, *11*, 1859.

[44] D. Imhof, U. Burckhardt, K.-H. Dahmen, F. Joho, R. Nesper, *Inorg. Chem.* **1997**, *36*, 1813.

[45] R. Dagani, *Chem. and Eng. News*, **1996**, January 1, 24.

[46] N. Feeder, J. Geng, P. G. Goh, B. F. G. Johnson, C. M. Martin, D. S. Shephard, W. Zhou, *Angew. Chem. Int. Ed.*, **2000**, *39*, 1661.

[47] E. Alonso, D. Astruc, *J. Am. Chem. Soc.,* **2000**, *122*, 3222.

[48] G. Schmid, W. Meyer-Zaika, R. Pugin, T. Sawitowski, J.-P. Majoral, A.-M. Caminade, C.-O. Turrin, *Chem. Eur. J.*, **2000**, *6*, 1693.

[49] N. T. Lucas, M. P. Cifuentes, L. T. Nguyen, M. G. Humphrey, *J. Cluster Sci.*, **2001**, *12*, 201.

[50] S. C. Tenhaeff, D. R. Tyler, *Organometallics* **1991**, *10*, 473.

[51] S. C. Tenhaeff, D. R. Tyler, *Organometallics* **1991**, *10*, 1116.

[52] S. C. Tenhaeff, D. R. Tyler, *Organometallics* **1992**, *11*, 1466.

[53] A. Avey, D. R. Tyler, *Organometallics* **1992**, *11*, 3856.

[54] D. R. Tyler, J. J. Wolcott, G. F. Nieckarz, S. C. Tenhaeff, In *"Inorganic and Organometallic Polymers II: Advanced Materials and Intermediates"*, P. Wisian-Neilson, H. R. Allcock, K. J. Wynne, Eds., American Chemical Society, Washington DC 1994, Vol. 572.

[55] G. F. Nieckarz, D. R. Tyler, *Inorg. Chim. Acta* **1996**, *242*, 303.

[56] G. F. Nieckarz, J. J. Litty, D. R. Tyler, *J. Organomet. Chem.* **1998**, *554*, 19.

[57] N. T. Lucas, M. G. Humphrey, A. D. Rae, *Macromolecules* **2001**, *34*, 6188.

[58] N. T. Lucas, E. G. A. Notaras, M. P. Cifuentes, M. G. Humphrey, *Organometallics* **2003**, *22*, 284.

[59] N. T. Lucas, E. G. A. Notaras, S. Petrie, R. Stranger, M. G. Humphrey, *Organometallics* **2003**, *22*, 708.

[60] E. G. A. Notaras, N. T. Lucas, M. G. Humphrey, A. C. Willis, A. D. Rae, *Organometallics* **2003**, *22*, 3659.

[61] N. T. Lucas, E. G. A. Notaras, M. G. Humphrey, M. Samoc, B. Luther-Davies, *SPIE Proc., Int. Soc. Opt. Eng.* **2003**, *5212*, 318.

Supramolecular Architectures Featuring Stereoisomeric Cluster Complexes of the $[Re_6(\mu_3\text{-}Se)_8]^{2+}$ Core

Hugh D. Selby, Bryan K. Roland, Jenine R. Cole, Zhiping Zheng*

Department of Chemistry, University of Arizona, Tucson, AZ 85721, USA

E-mail: zhiping@u.arizona.edu

Summary: The $[Re_6(\mu_3\text{-}Se)_8]^{2+}$ core-containing cluster complexes of the general formula $[Re_6(\mu_3\text{-}Se)_8(PEt_3)_4L_2]^{2+}$ (both *trans-* and *cis*-isomers) site-differentiated with inert PEt_3 and functional L ligands that are capable of hydrogen bonding or secondary (with respect to primary coordination with the cluster core) metal-ligand coordination interactions have been prepared. The applications of such stereospecific cluster isomers as building blocks for supramolecular construction have been studied. A great variety of multicluster arrays mediated by intercluster hydrogen bonding or cluster ligand coordination with secondary metal ions have been obtained and structurally characterized. The findings from this research clearly establish the superior utility of these novel building blocks for the creation of structurally sophisticated architectures and possibly functional materials

Keywords: hydrogen bonding; metal clusters; metal coordination; self-assembly; supramolecular structures

Introduction

Supramolecular chemistry, the "chemistry beyond the molecule", is based on the notion of creating novel structural and functional extended systems by linking prefabricated molecular or ionic building blocks via intermolecular interactions.[1] The designing principles, i.e. the intermolecular forces, are relatively well understood, and extensive research in the past few decades has produced numerous supramolecular constructs, many of which are not only of consummate structural beauty, but also have interesting and potentially useful properties.[2] Although there are certainly underlying principles and design rules yet to be discovered, further development in the future of this interdisciplinary research field appears to hinge upon our creativity to design novel building blocks in order to manipulate and utilize the known intermolecular forces.

DOI: 10.1002/masy.200450502

24

In light of this observation, over the past few years we have focused our efforts on the development of supramolecular chemistry utilizing metal clusters as building blocks. From a structural viewpoint, the multiple metal sites available in a cluster allow for site-differentiation, that is, selective binding to purpose-specific ligands. Thus, a range of building blocks with systematically varied and rigidly fixed stereochemistry may be realized. The fixed stereochemistry imparts the shape and directionality critical to supramolecular synthesis. From the perspective of creating novel functional materials, metal clusters are attractive because, in addition to the anticipated magnetic, electronic, optical, or catalytic properties inherent to metal complexes, clusters frequently exhibit interesting traits that are unique to metal-metal bonded species.[3] Thus, the efforts to build cluster-supported supramolecular structures are expected to offer many fascinating research problems with potentially significant ramifications.

We have concentrated our efforts on the use as building blocks of hexarhenium chalcogenide clusters featuring the $[Re_6(\mu_3\text{-}Q)_8]^{2+}$ (Q = S, Se) core (Figure 1).[4] The research is not only stimulated by the potential applications of metal-chalcogenide clusters for heterogeneous catalysis, photovoltaics, and semiconductor fabrication,[5] it is also conveniently built on their extensive and flexible synthetic chemistry.[6] In this mini-review, the highlights from our efforts will be summarized. In the first section we will elaborate on why these clusters, more specifically those containing the $[Re_6(\mu_3\text{-}Se)_8]^{2+}$ core, are of special interest in supramolecular construction. We will then illustrate the use of various stereoisomers derived from the common cluster core for the creation of a great variety of supramolecular architectures via either intercluster hydrogen bonding interactions or secondary (with respect to the primary ligand-cluster bonding) metal coordination; cluster complexes serve as ligands for the coordination of single metal ions in the latter. We will conclude with some perspectives in this exciting and rapidly growing research area.

The $[Re_6(\mu_3\text{-}Q)_8]^{2+}$ Clusters: Synthetic and Physical Fundamentals

Soluble molecular clusters of the hexarhenium chalcogenide core are a rather recent development in cluster chemistry.[12] The 24-electron face-capped hexanuclear core can be viewed as an octahedron of rhenium atoms enclosed in a cube formed by substitutionally inert chalcogenide ligands. Using the dimensional reduction protocol,[7] halide-terminated clusters of the general formula $[Re_6(\mu_3\text{-}Q)_8T_6]^{4-}$ (T = Cl, Br, I; Q = S, Se) are obtained from the initial solid-state synthesis.[4c]

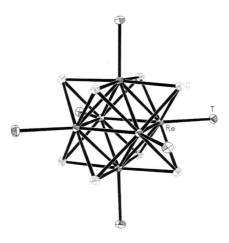

Figure 1. The structure of the $[Re_6(\mu_3\text{-}Q)_8]^{2+}$ cluster shown with terminal ligands (T).

At the heart of the $[Re_6(\mu_3\text{-}Q)_8]^{2+}$ (Q = S, Se) system's success is an extremely convenient derivatization (ligand substitution), affording a set of substitutional isomers of the cluster complexes.[6] Specifically, Holm and coworkers found that the six Re(III) sites could be differentiated from one another by reacting the starting $(n\text{-}Bu_4N)_3[Re^{III}{}_5Re^{IV}(\mu_3\text{-}Q)_8I_6]$ with varying ratios of PEt$_3$.[6a,b] Depending on the relative ratio of cluster to phosphine, the number of apical halides displaced could be varied from three to six. This was completely unlike any other known molecular cluster system. Instead of cluster decomposition or persubstitution, the phosphine substitution reactions could be executed in a controllable fashion. Citing biological terminology, the process was referred to as site differentiation.

Site differentiation was extremely significant however, as it marked the departure from cluster chemistry in the traditional sense. The phosphine ligands were completely inert once bound to the cluster core, thus fixing the stereochemistry of the system. More importantly, the remaining sites could be dehalogenated conveniently to afford the labile cluster solvates $[Re_6(\mu_3\text{-}Se)_8(PEt_3)_{6-n}(solv)_n]^{2+}$ (solv = MeCN, DMF, DMSO, and pyridine).[6a,8] These solvate sites were subject to further ligand exchange, making the structural possibilities limited only by imagination. The combination of easily controlled ligand exchange, robust cluster core, and large size allowed one to contemplate the synthesis of cluster complexes for more sophisticated purposes than elucidation of fundamental properties, for example, as a basis set of stable, chemically accessible, geometric building blocks for supramolecular construction. Although the use of mononuclear complexes in this role was already commonplace, no cluster system had ever been available in sufficient yields, with sufficient stability and range of stereochemistry, to replace them. With the site differentiated solvates in hand, it was clear that their large size and rigid stereochemistry would make them ideal components of supramolecular constructs.

Beyond the desirable structural characteristics, the extension of the $[Re_6(\mu_3\text{-}Se)_8]^{2+}$ system to endeavors not normally associated with clusters is supported by a host of potentially useful physical properties. The first of these is the now well characterized luminescence of all $[Re_6(\mu_3\text{-}Q)_8]^{2+}$ cores. Although initially the subject of some debate, it has been conclusively demonstrated that the cluster cores are luminescence. This was predicted by early computational work by Arratia-Perez and coworkers,[9a] and recently, established experimentally by Holm, Nocera, and their coworkers.[9b]

The $[Re_6(\mu_3\text{-}Q)_8]^{2+}$ cluster systems also feature extensive electrochemistry.[6] Like the luminescence, the electrochemistry is acutely sensitive to both inner and apical ligands. Incorporation of electronically bistable cluster complexes into supramolecular systems could result in novel electroactive materials for use in displays and other devices.[10]

Site-differentiated Cluster Complexes – A Set of Stereoisomeric Building Blocks

As alluded to in the above section, site-differentiation, the practice of protecting specific numbers of Re(III) coordination sites with inert phosphine ligands, affords a range of substitutional isomers. The stereochemistry of these isomers forms the geometric basis set of building blocks for construction of cluster-based supramolecules, provided that certain types of functional groups capable of secondary (with respect to primary cluster ligation) intermolecular interactions such as hydrogen bonding and metal-ligand coordination are integrated into the non-phosphine ligand(s). In this way, multicluster arrays mediated by the secondary interactions can be fabricated (Figure 2).

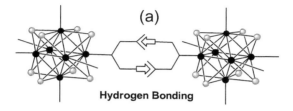

(a)

Hydrogen Bonding

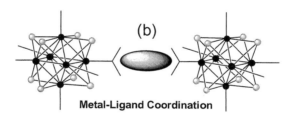

(b)

Metal-Ligand Coordination

Figure 2. Schematic demonstration of hydrogen bonding (a) and metal-ligand coordination (b) interactions for mediating the assembly of multicluster arrays.

As will be elaborated fully in the following sections, the fixed stereochemistry of the phosphine-protected clusters acts as the primary structure-directing element in the

28

synthetic process. Moreover, the wider displacement of ligands relative to their mononuclear counterparts minimizes structural ambiguity due to mononuclear complex flexibility.[11] Consequently *structural predictability* will be greatly enhanced. Furthermore, both the intercluster hydrogen bonding and the "cluster complex-as-ligand" approaches maintain the ease of handling offered by the monocluster species in solution yet offer the possibility of obtaining crystalline samples, frequently in the form of single crystals, of the supramolecular materials when assembled in the solid state.

Functional site-differentiated cluster complexes have thus been designed and synthesized (Scheme 1). In the first (**2A**, **2B**, **4A**, and **4B**), a complex is formed with ligands (isonicotinamide, for example) that bear functional groups capable of intermolecular hydrogen bonding. In the second (**2C** and **4C**), a cluster complex features ligands (4,4'-dipyridyl, for instance) that possess free coordinating atoms potentially exploitable for secondary metal coordination. The corresponding stereospecific acetonitrile solvates (**1** and **3**) can be readily prepared by deiodination of the respective iodo complexes with $AgSbF_6$ in the presence of MeCN.

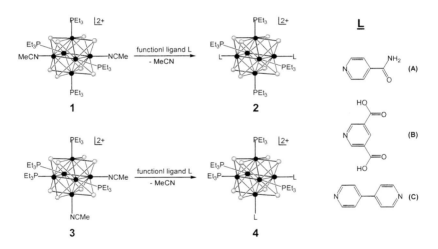

Scheme 1. Syntheses of site-differentiated $[Re_6(\mu_3\text{-Se})_8]^{2+}$ cluster complexes featuring inert PEt_3 ligands and ligands (L) capable of either hydrogen bonding (**A** and **B**) or ligand-metal coordination (**C**) interactions.

The Hydrogen Bonding Approach to Supramolecular Architectures

The hydrogen bond is not only one of the strongest intermolecular forces but has the added advantage of being directional. By using the site-differentiated cluster complexes as the hydrogen-bonding basis, the geometric and dimensional control of the cluster is added to the directionality of the hydrogen bond. We have prepared a number of site-differentiated cluster complexes with hydrogen bonding-capable ligands.[12-14] The solid-state structures of several of these clusters have been established crystallographically, revealing aesthetically pleasing and rather sophisticated hydrogen bonded supramolecular arrays of cluster building blocks. As an example, the supramolecular structure of *trans*-[$Re_6(\mu_3$-Se$)_8$(PEt$_3)_4$(isonicotinamide)$_2$](SbF$_6)_2$ (**2A**, Scheme 1),[12] each amide unit engages in a pair of self-complementary hydrogen bonding to two neighboring clusters, generating an infinite, one-dimensional chain featuring the cluster units and the cluster-linking hydrogen bonds (a, Figure 3).

The *cis* cognate of **2A**, *cis*-[$Re_6(\mu_3$-Se$)_8$(PEt$_3)_4$(isonicotinamide)$_2$]$^{2+}$ (**4A**, Scheme 1),[12] also features one-dimensional infinite chains of clusters in the solid state, which display a zigzag topology due to *cis*- displacement of the isonicotinamide ligands (b, Figure 3). This zigzag arrangement results in the chains fitting together to form sheets of chains (not shown). The lamellae are apparently stabilized by hydrophobic interdigitation of the phosphine groups between layers. The formation of a hydrogen-bonded square was not realized, probably due to the otherwise thermodynamically disfavored porous structure and packing thereof.

Using *fac*-[$Re_6(\mu_3$-Se$)_8$(PPh$_3)_3$(isonicotinamide)$_3$](SbF$_6)_2$,[13] prepared by reacting *fac*-[$Re_6(\mu_3$-Se$)_8$(PPh$_3)_3$(MeCN)$_3$](SbF$_6)_2$ with excess isonicotinamide, attempts have also been made to generate a discrete, hydrogen-bonded cluster cube, featuring eight corner-occupying cluster units and twelve edges of paired hydrogen bonds. Instead of a cube however, the cluster complexes form three-dimensional channels by packing such that the isonicotinamide cluster faces form the inner walls of the channels (top, Figure 4). The channels are stitched together by weaker, non-complementary hydrogen bonds from the "equatorial" amides in adjacent clusters along the channel axis (bottom, Figure 4). The very bulky and hydrophobic PPh$_3$ cluster faces form the outer walls. The structure may be stabilized by the clear separation of the relatively polar hydrogen bonding faces from the hydrophobic PPh$_3$, which subsequently engage in extensive π-π interactions with neighboring channels.

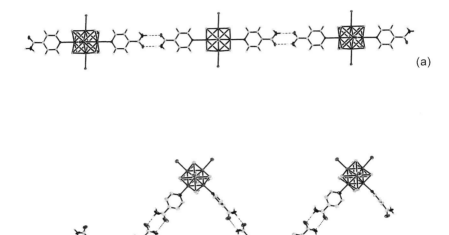

(a)

(b)

Figure 3. ORTEP views (50% ellipsoids) of hydrogen-bonded arrays of the [Re$_6$(μ_3-Se)$_8$]$^{2+}$ clusters [a(**2A**) and b(**4A**)].

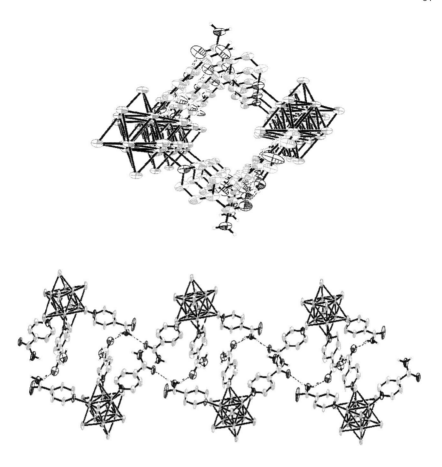

Figure 4. An ORTEP view (50% ellipsoids) of the extended hydrogen-bonded structure of fac-[Re$_6$(μ_3-Se)$_8$(PPh$_3$)$_3$(isonicotinamide)$_3$](SbF$_6$)$_2$. Triphenylphosphine ligands are removed for clarity.

Encouraged by the success in generating hydrogen bonded-supramolecular cluster arrays, cluster complexes **2B** and **4B** (Scheme 1) were prepared, in efforts to explore new modes of generating hydrogen bonded networks supported by cluster stereoisomers.[14] The non-phosphine ligand, namely 3,5-pyridinedicarboxylic acid (**B**), bares two hydrogen bonding donor (D) and acceptor (A) groups placed at 120E from each other on the pyridyl ring. The DA groups are subject to a variety of different hydrogen bonding modes. The combined effect of the increased number of hydrogen bonding DA units per molecule, and the increased variability of the hydrogen bonding modes, is to reduce the influence of the cluster stereochemistry over the final self-assembled architecture. Thus, the resultant hydrogen bonded arrays should reflect a synergy between cluster and ligand geometry, possibly affording novel structure and function not possible with only one component dominating the self-assembly process.

The solid-state structure and supramolecular arrangement of *trans*-[Re$_6$(μ_3-Se)$_8$(PEt$_3$)$_4$(3,5-pyridinedicarboxylic acid)$_2$](SbF$_6$)$_2$ (**2B**) have been established by crystallographic analysis (Figure 5).[14] The centroid of the cluster sits on an inversion center. The unique 3,5-pyridinedicarboxylic acid ligand is bound to a Re(III) center via the pyridyl nitrogen atom. Applying the inversion operator reveals that the 3,5-pyridinedicarboxylic acid ligands, including the carboxylic acid moieties, are virtually coplanar with the plane of four Re atoms that includes the two to which they are bound. The inversion related carboxylic acid groups undergo complementary hydrogen bonding with the neighboring sets. Interestingly, the second acid group on each ligand does not participate in any hydrogen bonding. Instead, the acid group extends into a small space between the chains and appears to be in close contact with a selenium atom on a neighboring cluster, as well as a hydrogen atom of that cluster's PEt$_3$ ligands. As the ligands are *trans*-coordinated to Re(III) sites, and only one of the two sets per ligand engage in hydrogen bonding, these carboxylic acid moieties may be considered *trans*-related across the complex. The result of this *trans-ligand, trans-acid* arrangement is the formation of zigzag hydrogen bonded cluster polymers.

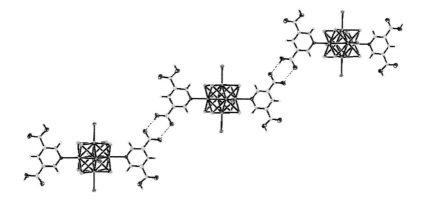

Figure 5. An ORTEP view (50% ellipsoids) of the extended hydrogen-bonded structure of *trans*-[Re$_6$(μ_3-Se)$_8$(PEt$_3$)$_4$(3,5-pyridinedicarboxylic acid)$_2$](SbF$_6$)$_2$ (**2B**, Scheme 1). Ethyl groups of the triphenylphosphine ligands are removed for clarity.

The "Cluster Complex-as-Ligand" Approach to Supramolecular Architectures

Using cluster complexes as ligands to direct secondary metal ion coordination represents the second half of our noncovalent cluster assembly paradigm. With the demonstration of the synthesis of hydrogen bonded arrays directed and supported by the [Re$_6$(μ_3-Se)$_8$]$^{2+}$ cluster stereoisomers, we were confident that the cluster core would be ideal in the role of "cluster complex-as-ligand". As compared with traditional organic ligands, complex ligands offer two distinct advantages. The first and most obvious is the that a complex ligand will bring its own physicochemical properties to the ensemble, offering the tantalizing prospect of creating materials with desirable properties such as magnetic, catalytic, and photo- and electrochemical activity. Second, the geometric preferences of a given complex ligand allow the realization of building blocks with shapes not trivially nor cheaply generated with purely organic molecules. As such, more sophisticated structures may be anticipated relative to traditional coordination polymers and oligomers. Admittedly, the coordination to a naked metal ion sacrifices some of the directionality and predictability of a hydrogen bonded linkage, the large size and rigid displacement of cluster-supported ligands are likely to ameliorate some of the potentially complicating factors associated with self-assembly (e.g. packing forces, ligand flexibility).

As a proof-of-concept first step, we synthesized $trans$-[Re$_6$(μ_3-Se)$_8$(PEt$_3$)$_4$(4,4'-dipyridyl)$_2$](SbF$_6$)$_2$ (**2C**, Scheme 1) and studied its coordination with a number of transition metal ions.[15] This cluster complex features two $trans$- disposed 4,4'-dipyridyl ligands; each of the pyridyl ligands coordinates the cluster core with one N atom, while leaving the other available for potential coordination to secondary metal ion. In effect, the complex is an expanded pyridyl-based ditopic ligand that can be used in lieu of purely organic equivalents to mediate the self-assembly of extended metal ion arrays.

Isomorphous linear coordination polymers based on $trans$-[Re$_6$(μ_3-Se)$_8$(PEt$_3$)$_4$(4,4'-dipyridyl)$_2$](SbF$_6$)$_2$ and Co(NO$_3$)$_2$ (**2C·Co**) and Cd(NO$_3$)$_2$ (**2C·Cd**) have been obtained and structurally characterized. Compound **2C·Co** is the first complex synthesized in the series of linear polymers (Figure 6). It features a repeat unit consisting of a single $trans$-[Re$_6$(μ_3-Se)$_8$(PEt$_3$)$_4$(4,4'-dipyridyl)$_2$] unit bound to a Co^{2+} ion via the open nitrogen of a single 4,4'-dipyridyl ligand. The Co^{2+} ion is bound to a second pyridyl nitrogen from the next repeat unit. Charge is balanced by a single SbF$_6^-$ counterion located near the cluster and three nitrato ligands bound to the Co^{2+} ion. The polymer formed by this repeat unit has significant curvature between the Re atom coordinated to one end of the 4,4'-dipyridyl ligand and the Co^{2+} unit at the other. The curvature is the result of the summation of small shifts from linearity at each individual bond between the two metals, and is a feature shared with its Cd^{2+} cognate. The curvature results in the polymers forming sinusoidal chains of modest amplitude. Adjacent chains have the cluster and cadmium units shifted with respect to one another, leading to alternating layers of opposing "phases" and an overall lamellar structure in the solid state with each layer composed of parallel polymer chains.

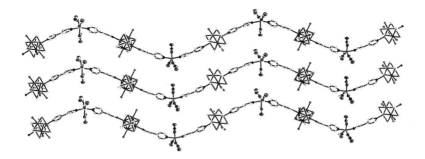

Figure 6. One layer of the Co^{2+}-mediated cluster assembly **2C·Co** shown with ethyl groups, hydrogen atoms, unbound counterions, and solvent omitted for clarity.

Using Zn^{2+} as the secondary metal ion, a chemically very similar coordination polymer (**2C·Zn**) was obtained. Structurally however, it is quite distinct (Figure 7). The repeat unit is essentially the same as **2C·Co** and **2C·Cd** except that the $N_{pyridyl}$-Zn^{2+}-$N_{pyridyl}$ coordination mode is *cis*- with respect to the Zn^{2+} ion, with an average $N_{pyridyl}$-Zn^{2+}-$N_{pyridyl}$ angle of 88<. The effect of this coordination mode is to produce pseudo-linear zigzag polymer chains, rather than the sinusoidal polymers of **2C·Co** and **2C·Cd**. Indeed, one might consider the zigzag motif the canonical limit of the curvature observed in the previous structures: The relatively rigid 4,4'-dipyridyl ligand can only accommodate a certain amount of distortion before adopting the *cis*- configuration about the secondary metal ion. The relative contributions of electronic, steric, and packing energetics that result in a given configuration remain speculative, however. As in the previous two structures, the zigzag polymers form layers of parallel chains, but the layers are interpenetrated. In the spaces between chains in a given sheet, chains forming the penetrating layer run roughly normal to the plane of the sheet. This packing mode results in the formation of small channels that pass through the crystal in the *b* direction between neighboring Zn^{2+} sites (not shown).

Figure 7. A single chain of the Zn^{2+}-mediated cluster assembly (**2C·Zn**). Ethyl groups, protons, counterions, and solvent omitted for clarity.

The linear polymers **2C·Co**, **2C·Cd**, and **2C·Zn** are aesthetically pleasing supramolecules and serve as excellent proofs-of-concept for the utility of cluster expanded "big-bpy" type ligands. The second motif was inspired by our original work with molecular squares[16] and was intended to increase structural and functional sophistication in a rational way. By using cis-[Re$_6$(μ_3-Se)$_8$(PPh$_3$)$_4$(4,4'-dipyridyl)$_2$](SbF$_6$)$_2$ (**4C**, Scheme 1), a ligand equipped with an enforced right angle between the two dipyridyl ligands, we hoped to realize a very large molecular square with **4C** forming the corners.

An unexpected product was produced upon ether diffusion into a 1:1 (molar) mixture of **4C** in dichloromethane and Cd(NO$_3$)$_2$ in methanol. As revealed by crystallographic analysis, the product, formulated as {[Re$_6$(μ_3-Se)$_8$(PPh$_3$)$_4$(4,4'-dipyridyl)$_2$]$_2$[Cd(NO$_3$)$_2$]}(SbF$_6$)$_4$ (**4C·Cd1**) exhibits a 2:1 cluster/Cd^{2+} ratio and forms a one-dimensional chain of corner-sharing squares (left, Figure 8) in the solid state.[17] Each shared corner is a single octahedrally coordinated Cd^{2+} ion, bound in the equatorial plane by four pyridyl nitrogens and axially by two η^1-NO$_3^-$ ligands. The unshared corners are formed by the [Re$_6$(μ_3-Se)$_8$]$^{2+}$ clusters.

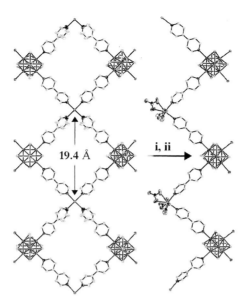

19.4 Å

i, ii

Figure 8. Thermal ellipsoid plots of **4C•Cd1** (left) and **4C•Cd2** (right) rendered at 50% probability. Only the framework structure is shown. Conversion in two steps: (i) Cd(NO$_3$)$_2$/CH$_3$OH; (ii) ether diffusion.

A complex of a 1:1 cluster/Cd^{2+} ratio was subsequently attempted and obtained by reacting **4C** with a large excess of the Cd^{2+} salt. The product, formulated as [{Re$_6$(μ_3-Se)$_8$(PPh$_3$)$_4$(4,4'-dipyridyl)$_2$}{Cd(NO$_3$)$_3$}](NO$_3$) (**4C•Cd2**) is a one-dimensional zigzag chain in the solid state (right, Figure 8). As with **4C•Cd1**, the chains clearly reflect the geometric constraint imposed by the cluster ligands, but also show flexibility at the Cd^{2+} site. The coordination environment about the Cd^{2+} ion is distorted trigonal bipyramidal, with one pyridyl axial and one equatorial (N-Cd-N bond angle of 87.2°). The remaining three sites are occupied by η^2-NO$_3^-$ ligands. Though disappointing, the outcome was not too surprising as the polymeric structure is indeed the more likely product when considered from a crystal energetics perspective.

38

Curiously, the chain structure of **4C•Cd2** can be formally considered to be derived from **4C•Cd1** by replacing the cluster on one side of the squares with a nitrato ligand. Indeed, as confirmed by crystallographic determination of cell parameters, complex **4C•Cd2** was isolated from the reaction of complex **4C•Cd1** with an excess of $Cd(NO_3)_2$ in a methanolic solution upon ether vapor diffusion. However, the extended structure of **4C•Cd1** is presumably disintegrated in a highly polar solvent like methanol into the component cluster complex and Cd^{2+} prior to their re-assembly, with the additional $Cd(NO_3)_2$, into the zigzag chain of **4C•Cd2**.

Conclusions

The purpose of the research is the development of synthetic methodologies necessary to bring the $[Re_6(\mu_3\text{-Se})_8]^{2+}$ system out of the limited sphere of fundamental cluster chemistry and into general synthetic applicability. The significance of this research is born of its novelty – but not the trite observation that no one has done such an investigation before. Rather, the research is significant because no other inorganic system, cluster or otherwise, exhibits the same range of reactivity, durability, and physicochemical activity that makes the $[Re_6(\mu_3\text{-Q})_8]^{2+}$ system so useful. Specifically, we have explored the uses of site-differentiated cluster complexes featuring secondary functionality placed at the ends of pyridyl based-ligands as expanded and stereospecific building blocks for supramolecular construction. These secondary functions are moieties capable of hydrogen bonding or metal ion coordination. The cluster's versatility in this role is manifested in the variety of beautiful architectures obtained and the facile change of synthetic paradigms necessary to achieve these results. Although we have explored a range of structural types and their synthetic methods, often with interesting and unexpected results, we are only at the beginning of a very exciting journey. It is obvious that Nature still has a few tricks up in her sleeve, but the complete set of building blocks available from the cluster basis, combined with the adaptability to any assembly mode, will no doubt make the task of uncovering these secrets considerably easier.

Acknowledgements

We wish to thank Research Corporation and University of Arizona for financial support of this work. This work is also partially supported by Petroleum Research Fund administered by the American Chemical Society. The CCD-based X-ray diffractometer was purchased through an NSF grant (CHE-96103474, USA).

[1] J.-M. Lehn, *"Supramolecular Chemistry: Concepts and Perspectives"* VCH, New York 1995.
[2] For selected reviews, see: [2a] O. W. Evans, W. Lin, *Acc. Chem. Res.* **2002**, *35*, 511; [2b] B. Moulton, M. J. Zaworotko, *Chem. Rev.* **2001**, *101*, 1629; [2c] L. J. Prins, D. N. Reinhoudt, P. Timmerman, *Angew. Chem., Int. Ed.* **2001**, *40*, 2382; [2d] S. Leininger, B. Olenyuk, P. J. Stang, *Chem. Rev.* **2000**, *100*, 853; [2e] D. N. Caulder, K. N. Raymond, *J. Chem. Soc., Dalton Trans.* **1999**, 1185; [2f] O. M. Yaghi, H. Li, C. Davis, D. Richardson, T. L. Groy, *Acc. Chem. Res.* **1998**, *31*, 474; [2g] M. Fujita, *Chem. Soc. Rev.* **1998**, *27*, 417; [2h] S. R. Batten, R. Robson, *Angew. Chem., Int. Ed.* **1998**, *37*, 1460; [2i] M. C. T. Fyfe, J. F. Stoddart, *Acc. Chem. Res.* **1997**, *30*, 393.
[3] F. A. Cotton, C. Lin, C. A. Murillo, *Acc. Chem. Res.* **2001**, 34, 759.
[4] [4a] J. C. P. Gabriel, K. Boubekeur, S. Uriel, P. Batail, *Chem. Rev.* **2001**, *101*, 2037; [4b] T. Saito, *J. Chem. Soc., Dalton Trans.* **1999**, 97; [4c] J. R. Long, L. S. McCarty, R. H. Holm, *J. Am. Chem. Soc.* **1996**, *118*, 4603.
[5] F. J. Disalvo, *Annales de Chimie* **1982**, *7*, 109.
[6] [6a] Z. Zheng, J. R. Long, R. H. Holm, *J. Am. Chem. Soc.* **1997**, *119*, 2163; [6b] M. W. Willer, J. R. Long, C. C. McLauchlan, R. H. Holm, *Inorg. Chem.* **1998**, *37*, 328; [6c] Z.-N. Chen, T. Yoshimura, M. Abe, Y. Sasaki, S. Ishizaka, H.-B. Kim, N. Kitamura, *Angew. Chem., Int. Ed.* **2001**, *40*, 239.
[7] E. G. Tulsky, J. R. Long, *Chem. Mater.* **2001**, *13*, 1149.
[8] [8a] Z. Zheng, R. H. Holm, *Inorg, Chem.* **1997**, *36*, 5173; [8b] Z. Zheng, T. G. Gray, R. H. Holm, *Inorg. Chem.* **1999**, *38*, 4888.
[9] [9a] R. Arratia-Pérez, L. Hernández-Acevedo, *J. Chem. Phys.* **1999**, *111*, 168; [9b] T. G. Gray, C. M. Rudzinski, E. E. Meyer, R. H. Holm, D. G. Nocera, *J. Am. Chem. Soc.* **2003**, *125*, 4755.
[10] Köhler, J. S. Wilson, R. H. Friend, *Adv. Mater.* **2002**, *14*, 701.
[11] M. Schweiger, S. R. Seidel, A. M. Arif, P. J. Stang, *Angew. Chem., Int. Ed.* **2001**, *40*, 3467.
[12] H. D. Selby, B. K. Roland, M. D. Carducci, Z. Zheng, *Inorg. Chem.* **2003**, *42*, 1656.
[13] H. D. Selby, Z. Zheng, unpublished results.
[14] K. Roland, H. D. Selby, J. R. Cole, Z. Zheng, *Dalton Trans.* **2003**, 4307.
[15] H. D. Selby, P. Orto, Z. Zheng, *Polyhedron* **2003**, *22*, 2999.
[16] H. D. Selby, Z. Zheng, T. G. Gray, R. H. Holm, *Inorg. Chim. Acta* **2001**, *312*, 205.
[17] H. D. Selby, P. Orto, M. D. Carducci, Z. Zheng, *Inorg. Chem.* **2002**, *41*, 6175.

Artificial Metallo-DNA towards Discrete Metal Arrays

Mitsuhiko Shionoya

Department of Chemistry, Graduate School of Science, The University of Tokyo, Hongo, Bunkyo-ku, Tokyo 113-0033, Japan
E-mail: shionoya@chem.s.u-tokyo.ac.jp

Summary: DNA shows promise as a provider of a structural basis for the "bottom-up" fabrication of functionalized molecular building blocks. In particular, the replacement of hydrogen-bonded DNA base pairing for alternative one could possibly provide a novel tool for re-engineering DNA as well as for biological applications. This review describes our recent approaches to metal-based strategy directed towards self-assembled metal arrays within DNAs. Recently, we reported the synthesis of a series of artificial oligonucleotides, d(5'-GH_nC-3') (n = 1-5), using hydroxypyridone nucleobases (***H***) as flat bidentate ligands. Right-handed double helices of the oligonucleotides, nCu^{2+}·d(5'-GH_nC-3')$_2$ (n = 1-5), are quantitatively formed through Cu^{2+}-mediated alternative base pairing (***H***-Cu^{2+}-***H***), where the Cu^{2+} ions are aligned along the helix axes inside the duplexes with the Cu^{2+}-Cu^{2+} distance of 3.7 ± 0.1 Å. The Cu^{2+} ions are coupled in a ferromagnetic manner with one another through unpaired d electrons to form magnetic chains. This strategy represents a new method for self-assembled metal arrays in a predesigned fashion, leading to the possibility of metal-based molecular devices such as molecular magnets and wires.

Keywords: artificial DNA; base pairing; metal array; metal complex; metallo-DNA; self-assembly

Introduction

Research on bio-inspired molecular architecture is often directed towards the redesign of fundamental building blocks that have been provided by Nature and then the "bottom-up" syntheses of a wide range of possible structures and functions. Although biomacromolecules contain only a limited number of building blocks such as nucleotides and amino acids, owing to recent advances in chemical synthesis and biotechnology, one can replace the building blocks by chemically modified ones to arrange them one after another with a desired length and sequence.

In addition, self-assembly protocols and template-directed procedures, which are efficiently used in the biological systems, have been conceptually introduced into chemical approaches to self-assembled, nano-sized molecules or materials. Herein we focus on the incorporation of metal complexes as alternative components into biomolecular scaffolds, which is a key design in the structural control and functionalization of biopolymers.

DNA provides a structural basis to arrange functionalized molecular building blocks into predesigned geometries. In the double-stranded DNA, hydrogen-bonded base pairs, which are attached nearly perpendicular to the phosphate backbone, are arranged into direct stacked contact. Therefore, among a variety of approaches to DNA-based supramolecular architectures, the strategy of replacing natural DNA base pairs by alternative ones possessing a distinctive shape, size, and function[1] would be expected to provide a general method of molecular arrangement within the DNA in a controllable manner. Such extra base pairs would not only expand the genetic alphabet but would also allow the replication of DNA containing unique functional groups. Moreover, DNA that is completely built out of artificial base pairs could lead to novel oligomers or polymers having unique chemical and physical properties. This review describes recent advances in artificial metallo-DNAs directed toward DNA nanotechnology as well as gene control.

Alternative Hydrogen Bonding and Non-Hydrogen Bonding Schemes for DNA Base Pairing

Watson-Crick hydrogen bonding in natural base pairs plays crucial roles in the DNA functions. Initial effort was made mainly to expand the genetic alphabet using altered hydrogen bonding. Benner and co-workers pioneered an excellent way to enzymatic incorporation of new hydrogen-bonded base pairs into DNA/RNA to extend the genetic alphabet.[2a] They reported a series of nucleobase analogs whose hydrogen-bonded patterns differ from those in the adenine-thymine (A-T) and guanine-cytosine (G-C) base pairs of natural DNA. Interestingly, even subtle changes in their hydrogen-bonding patterns have a great influence on thermodynamic[2b] and biochemical properties.[2c-e]

Watson-Crick base pairing in DNA keeps two rules of complementarity in both size and hydrogen bonding patterns. Hydrophobicity and planarity of the bases are also important for the stability of the double helical structure. In this context, Kool and co-workers have developed a series of shape mimics of natural bases lacking hydrogen bonding functionality, using the principle that two bases should be complementary in shape rather than in hydrogen bonding.[3] A set of non-hydrogen-bonding base mimics for thymine and adenine was designed based on the criteria that oxygen and nitrogen could be replaced by fluorine and carbon, respectively, keeping aromaticity intact. For example, the nucleoside bearing a difluorotoluene nucleobase is an excellent mimic of thymidine that can pair with adenine within DNAs.[3d,e] This out performing shape mimic of thymidine can also effectively substitute for thymine as the incoming substrate in the triphosphate form[4] as well as in the template strand[5a] in polymerase-related enzymatic reactions. These results overall suggest that enzymatic replication of base pairs does not need Watson-Crick hydrogen bonds as long as the components stack strongly and that shape recognition is important in the base pairing without hydrogen bonding.[5b] Others have also reported hydrophobic unnatural base pairs as attractive candidates for expansion of the genetic alphabets.[6]

Metal-Mediated Base Pairing in DNA

Basic Concept

When natural DNA bases are replaced by alternative ones that have the ability to bind metal ions, metal-mediated base pairs would be incorporated into DNAs at desired positions or could possibly be aligned along the helix axis. Since metal coordinative bond energy is intermediate between covalent and noncovalent ones, one metal-ligand bond should compensate for two or three hydrogen bonds as seen in the natural DNA base pairs. Metal ions thus incorporated would 1) stabilize high-order structures of DNA (duplex, triplex, etc), 2) allow one-dimensional metal arrays along the DNA helix axis with unique chemical and physical properties, 3) generate metal-dependent electro- or photochemical functions, 4) assemble DNA duplexes at the junctions to form two- or three-dimensional DNA networks, 5) label DNA at desired positions, and so on. Importantly, when metals are aligned into direct stacked contact within DNA duplexes, the net charge of each metal-assisted base pair needs to be controlled so that the electrostatic repulsion between positively charged metal centers can be reduced – that is, negatively charged bases should be suitable for metal arrays.

Artificial Nucleosides Designed for Metal-Mediated Base Pairs

Recently, we have reported the first metal-assisted base pair using a β-C-nucleoside having a phenylenediamine base as Pd^{2+}-mediated base pairing.[7] Since then, some other β-N- and β-C-nucleosides designed for metal-mediated base pairing have been reported. Examples of artificial nucleosides having mono- to tetradentate ligands for metals so far reported are shown in Figure 1. Each nucleobase has one to four donor atoms at proper positions so that the ligand moiety can form a 1:1 to 4:1 complex with a transition metal ion in a linear, trigonal-planar, square-planar, tetrahedral, or octahedral coordination geometry. Among these geometries, a square-planar, linear, or trigonal-planar metal complex is most likely to replace a flat, hydrogen-bonded natural base pair. So far, we have reported, in addition to the above mentioned Pd^{2+}-mediated base pairing, B^{3+}-induced base paring with catechol,[8] Pd^{2+}-mediated base pairing with 2-aminophenol,[9] Ag^+-assisted base pairing with pyridine,[10] and Cu^{2+}-mediated base pairing with hydroxypyridone[11] as alternative base pairing modes (Figure 2). Other groups have also reported metal-mediated pairing ligand nucleobase mimics such as a Cu^{2+}-mediated base pair between pyridine and pyridine-2,6-dicarboxylate,[12a,b] an Ag^+-mediated pairing with 2,6-bis(methylthiomethyl)pyridine bases,[12c] and "ligandosides" using bipyridine nucleobases (Figure 2).[13]

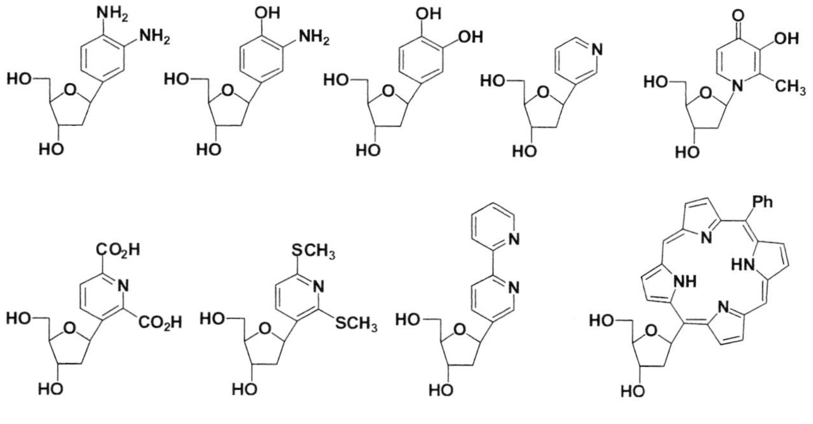

Figure 1. Examples of artificial nucleosides whose bases are replaced by metal ligands.

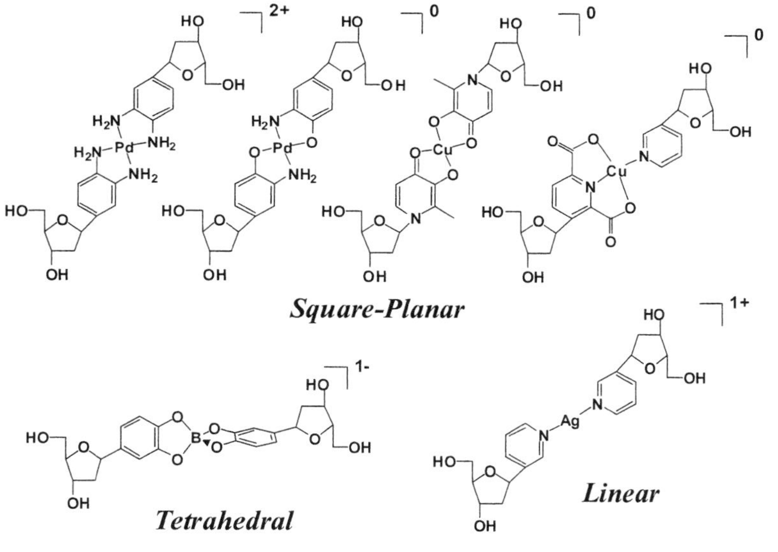

Square-Planar

Tetrahedral

Linear

Figure 2. Examples of metal-mediated base pairs through metal-coordinating nucleoside analogues.

Single-Site Incorporation of a Metallo-Base Pair into DNA

Artificial nucleobases can be incorporated into DNA using phosphoramidite derivatives of the nucleosides with standard protocols using an automated DNA synthesizer. The first example of metal-mediated base pairing in oligonucleotides was reported by Schultz and Romesberg et al.[12a] A set of a pyridine-2,6-dicarboxylate nucleobase as a planar tridentate ligand and a pyridine nucleobase as the complementary single donor ligand was incorporated into the middle of an oligonucleotide duplex. The duplex is significantly stabilized by the formation of a neutral Cu^{2+} complex with the paired ligand bases inside the DNA.

We have independently established single-site incorporation of an Ag^{+}-mediated base pair into a double-stranded DNA by introducing a monodentate pyridine nucleobase in the middle of each strand.[10] For example, an Ag^{+} ion incorporated into a DNA duplex, $d(5'-T_{10}PT_{10}-3')\cdot d(3'-A_{10}PA_{10}-5')$, containing a pyridine nucleobase (**P**) in the middle of the sequence increases the thermal stability of the duplex due to the formation of a positively charged **P**-Ag^{+}-**P** base pair. In contrast, this Ag^{+}-dependent thermal stabilization of duplex is only slight in a reference DNA duplex, $d(5'-T_{21}-3')\cdot d(3'-A_{21}-5')$. Considering the relatively weak binding between Ag^{+} and pyridine in aqueous media, it appears that the **P**-Ag^{+}-**P** base pairing is reinforced cooperatively by the surrounding hydrogen-bonded and stacked natural base pairs in the hydrophobic environment within the duplex. Moreover, this Ag^{+}–mediated base pairing is specific because the addition of other transition metal ions such as Cu^{2+}, Ni^{2+}, Pd^{2+}, and Hg^{2+} showed almost no significant effects on their melting processes of the duplexes. Such thermal stabilization as seen in the case of Ag^{+} is also observed with a triplex, $d(5'-T_{10}PT_{10}-3')\cdot d(3'-A_{10}PA_{10}-5')\cdot d(5'-T_{10}PT_{10}-3')$.[10] This effect is believed to be due to the formation of a Ag^{+}-mediated base triplet in which the three pyridine nitrogen donors from the three strands coordinate to the Ag^{+} center in a trigonal-planar coordination geometry.

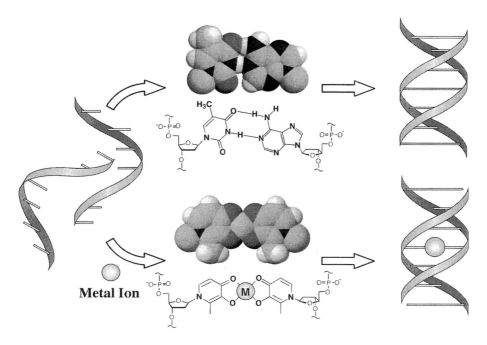

Figure 3. A schematic representation for single-site incorporation of a metal-mediated base pair.

A hydroxypiridone-bearing nucleoside was then incorporated into an oligonucleotide duplex, d(5'-CACATTA**H**TGTTGTA-3')·d(3'-GTGTAAT**H**ACAACAT-5'), to form a neutral Cu^{2+}-mediated base pair of hydroxypyridone nucleobases (**H**-Cu^{2+}-**H**) in the middle of the sequence (Figure 3).[11] In the presence of equimolar Cu^{2+} ions, an **H**-Cu^{2+}-**H** base pair is quantitatively formed within the DNA and the artificial duplex is more stabilized compared with a natural oligoduplex in which the **H**-**H** base pair is replaced by an **A**-**T** base pair. In addition, EPR and CD spectra of the metallo-DNA suggested that the radical site of a Cu^{2+} center is formed within the right-handed double-strand structure of the oligonucleotide. This strategy was further developed for controlled and periodic spacing of neutral metallo-base pairs along the helix axis of DNA.

Discrete Self-Assembled Metal Arrays in DNA

To control metal arrays at the molecular level in a discrete and predictable manner, one needs appropriate ligands having a varied number of coordination sites whose sequence can be programmable. If each metal binding site of the ligand is highly selective for some specific metal ion, the information of the sequence of the metal binding sites would be transferred into that of metals. From this standpoint, DNA, when the nucleobases are replaced by ligand-like bases, can be regarded as a multidentate ligand for one-dimensional metal arrays. In other words, the incorporation and the subsequent arrangement of metallo-base pairs into direct stacked contact within DNA could lead to "metallo-DNA" in which metal ions are lined up along the helix axis in a controlled manner.

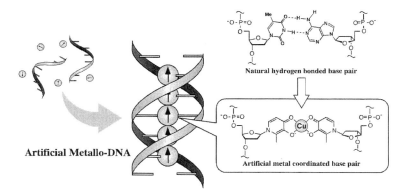

Figure 4. A schematic representation of Cu^{2+}-mediated duplex formation in which five Cu^{2+} ions are aligned along the helical axis within the DNA.

Recently, we reported the syntheses of a series of artificial oligonucleotides, $d(5'\text{-}GH_nC\text{-}3')$ ($n = 1\text{-}5$), using hydroxypyridone nucleobases (**H**) as flat bidentate ligands.[14] Photometric titration studies clealy showed that right-handed double helices of the oligonucleotides, $nCu^{2+} \cdot d(5'\text{-}GH_nC\text{-}3')_2$ ($n = 1\text{-}5$), are quantitatively formed through Cu^{2+}-mediated alternative base pairing (**H**-Cu^{2+}-**H**) (Figure 4). In these metallo-DNA, the Cu^{2+} ions incorporated into each complex are aligned along the helix axes inside the duplexes with the Cu^{2+}-Cu^{2+} distance of 3.7 ± 0.1 Å as evidenced by the EPR study. The Cu^{2+} ions are coupled with one another through unpaired *d* electrons to form magnetic chains. The electron spins on adjacent Cu^{2+} centers are aligned parallel and couple in a ferromagnetic fashion with accumulating Cu^{2+} ions attaining the highest spin state, exactly as expected from a line-up of Cu^{2+} ions. Such strategy could be developed for self-assembled metal arrays in a variety of DNA structures such as multi-stranded, hairpin, junction, or cyclic structures. Heterotopic metal arrays as well as elongation of metal strings by chemical and enzymatic methods is now underway.

Future Prospects for Artificial Metallo-DNA

Such a new binding motif in DNA duplex will influence research in areas as diverse as medicinal chemistry, materials science, and bio-nanotechnology. Introduction of metallo-base pairs into DNA would not only affect the stability of DNA double strands but also confer a variety of metal-based functions upon DNA. This strategy represents a new method for arranging metal ions in solution in a discrete and predictable fashion, leading to the possibility of metal-based molecular devices such as molecular magnets and wires. It is a real challenge to create hetero metal arrays with unique functions, leading possibly to chemical communication between different kinds of metals triggered by stimuli from outside.

[1]   [1a] D. E. Bergstrom, in: *"Current Protocols in Nucleic Acid Chemistry"*, S. L. Beaucage, D. E. Bergstrom, G. D. Glick, R. A. Jones, Eds., Wiley, New York 2001, Unit 1.4; [1b] E. T. Kool, *Acc. Chem. Res.* **2002**, *35*, 936.
[2]   [2a] J. A. Piccirilli, T. Krauch, S. E. Moroney, S. A. Benner, *Nature* **1990**, *343*, 33; [2b] J. J. Voegel, S. A. Benner, *J. Am. Chem. Soc.* **1994**, *116*, 6929; [2c] J. D. Bain, A. R. Chamberlin, C. Y. Switzer, S. A. Benner, *Nature* **1992**, *356*, 537; [2d] C. Y. Switzer, S. E. Moroney, S. A. Benner, *Biochemistry* **1993**, *32*, 10489; [2e] J. Horlacher, M. Hottiger, V. N. Podust, U. Hübscher, S. A. Benner, *Proc. Natl. Acad. Sci. U.S.A.* **1995**, *92*, 6329.
[3]   [3a] R. X.-F. Ren, N. C. Chaudhuri, P. L. Paris, S. Rumney IV, E. T. Kool, *J. Am. Chem. Soc.* **1996**, *118*, 7671; [3b] J. C. Morales, K. M. Guckian, C. Shails, E. T. Kool, *J. Org. Chem.* **1998**, *63*, 9652; [3c] B. M. O'Neill, J. E. Ratto. K. L. Good, D. Tahmassebi, S. A. Helquist, J. C. Morales, E. T. Kool, *J. Org. Chem.* **2002**, *67*, 5869; [3d] K. M. Guckian, E. T. Kool, *Angew. Chem., Int. Ed.* **1998**, *36*, 2825; [3e] K. M. Guckian, T. R. Krugh, E. T. Kool, *Nature Struct. Biol.* **1998**, *5*, 954; [3f] K. M. Guckian, T. R. Krugh, E. T. Kool, *J. Am. Chem. Soc.* **2000**, *122*, 6841; [3g] B. A. Schweizer, E. T. Kool, *J. Am. Chem. Soc.* **1995**, *117*, 1863.
[4]   S. Moran, R. X.-F. Ren, E. T. Kool, *Proc. Natl. Acad. Sci. U.S.A.* **1997**, *94*, 10506.
[5]   [5a] S. Moran, R,X.-F. Ren, S. I. Rumney, E. T. Kool, *J. Am. Chem. Soc.* **1997**, *119*, 2056; [5b] E. T. Kool, *Biopolymers (Nucleic Acid Sci.)* **1998**, *48*, 3.

50

[6]   [6a] D. E. Bergstrom, P. Zhang, P. H. Toma, C. A. Andrews, R. Nichols, *J. Am. Chem. Soc.* **1995**, *117*, 1201; [6b] P. Zhang, W. T. Johnson, D. Klewer, N. Paul, G. Hoops, V. J. Davisson, *Nulc. Acids Res.* **1998**, *26*, 2208; [6c] D. L. McMinn, A. K. Ogawa, Y. Wu, J. Liu, P. G. Schultz, F. E. Romesberg, *J. Am. Chem. Soc.* **1999**, *121*, 11585; [6d] M. Berger, A. K. Ogawa, D. L. McMinn, Y. Wu, P. G. Schultz, F. E. Romesberg, *Angew. Chem.*, *Int. Ed. Engl.* **2000**, *39*, 2940; [6e] A. K. Ogawa, Y. Wu, D. L. McMinn, J. Liu, P. G. Schultz, F. E. Romesberg, *J. Am. Chem. Soc.* **2000**, *122*, 3274; [6f] Y. Wu, A. K. Ogawa, M. Berger, D. L. McMinn, P. G. Schultz, F. E. Romesberg, *J. Am. Chem. Soc.* **2000**, *122*, 7621.

[7]   [7a] K. Tanaka, M. Shionoya, *J. Org. Chem.* **1999**, *64*, 5002; [7b] M. Shionoya, K. Tanaka, *Bull. Chem. Soc. Jpn.* (Account) **2000**, *73*, 1945.

[8]   H. Cao, K. Tanaka, M. Shionoya, *Chem. Pharm. Bull.* **2000**, *48*, 1745.

[9]   [9a] K. Tanaka, M. Tasaka, H. Cao, M. Shionoya, *Eur. J. Pharm. Sci.* **2001**, *13*, 77; [9b] M. Tasaka, K. Tanaka, M. Shiro, M. Shionoya, *Supramol. Chem.* **2001**, *13*, 671; [9c] K. Tanaka, M. Tasaka, H. Cao, M. Shionoya, *Supramol. Chem.* **2002**, *14*, 255.

[10] K. Tanaka, Y. Yamada, M. Shionoya, *J. Am. Chem. Soc.* **2002**, *124*, 8802.

[11] K. Tanaka, A. Tengeiji, T. Kato, N. Toyama, M. Shiro, M. Shionoya, *J. Am. Chem. Soc.* **2002**, *124*, 12494.

[12] [12a] E. Meggers, P. L. Holland, W. B. Tolman, F. E. Romesberg, P. G. Schultz, *J. Am. Chem. Soc.* **2000**, *122*, 10714; [12b] S. Atwell, E. Meggers, G. Spraggon, P. G. Schultz, *J. Am. Chem. Soc.* **2001**, *123*, 12364; [12c] N. Zimmermann, E. Meggers, P. G. Schultz, *J. Am. Chem. Soc.* **2002**, *124*, 13684.

[13] [13a] H. Weizman, Y. Tor, *Chem. Commun.* **2001**, 43; [13b] H. Weizman, Y. Tor, *J. Am. Chem. Soc.* **2001**, *123*, 3375.

[14] K. Tanaka, A. Tengeiji, T. Kato, N. Toyama, M. Shionoya, *Science* **2003**, *299*, 1212.

Macromol. Symp. **2004**, *209*, 51-65

Novel Carbazole Dendrimers Having a Metal Coordination Site as a Unique Hole-Transport Material

Atsushi Kimoto, Jun-Sang Cho, Masayoshi Higuchi, Kimihisa Yamamoto*

Kanagawa Academy of Science & Technology (KAST), and Department of Chemistry, Faculty of Science & Technology, Keio University, Yokohama, 223-8522 (Japan)
E-mail: yamamoto@chem.keio.ac.jp

Summary: We have synthesized novel carbazole dendrimers via the cyclotrimerization of aminophenylketones in the presence of titanium tetrachloride. These dendrimers have the ability to assemble metal ions such as Sn^{2+} and Eu^{3+} with no significant difference in their generation, suggesting the dendrimer with an interior with a small density up to the third generation. We show the dendrimers with higher generations have the higher HOMO values. The most electron rich molecule, the G3 dendrimer, has the highest HOMO value of -5.2 eV. However, for the HOMO energy levels of the carbazole dendrimer complex with $Eu(OTf)_3$, the energy levels of the carbazoles did not change based on almost the same redox potentials as those of the dendrimers, themselves. Using the carbazole dendrimers and their europium complex, a homogeneous film was produced, which enhanced the performance of the electroluminescence device in comparison with only the dendrimer as the hole-transporting layer. This approach was managed by a solution process, i.e., the spin-coating method, without using the coevaporation technique based on the large equilibrium constants of the coordination of metal ions on the imine sites ($K = 10^5 M^{-1}$).

Keywords: carbazole; dendrimers; electroluminescence; hole transport; metal-polymer complexes

Introduction

Discrete triphenylamine arrays based on linear,[1] cyclic,[2] and dendritic[3] systems with a conjugated backbone display attractive magnetic and electronic properties. Also, acting as hole-transport materials in various electro-optical applications, for example, organic light-emitting diodes (OLEDs)[4] and photocells,[5] branched and dendritic structures exhibit a better amorphous property and high solubility due to the geometry of these molecules without close packing. Above all, dendrimers consisting of a core, dendrons, and peripheral units can be easily designed

DOI: 10.1002/masy.200450504

by modular synthesis, however, hyper-branched polymers cannot. We should note that dendrimers, especially rigid ones, can possibly be regularly assembled by packing on a plate without deformation of the molecule[6] and are expected to expand the field of nanomaterials.[7] Organic materials for various electro-optical applications generally consist of rigid π-conjugated structures with a narrow HOMO-LUMO gap. Above all, a number of dendrimers have been used in OLEDs[8] designed by their characteristic synthetic procedure, the convergent method. The advantages of adopting monodisperse and well-defined dendrimers as active components in OLEDs are that they can be easily prepared in high purity and have a better amorphous property and high solubility due to their geometry without close packing, resulting in the easier fabrication of thin films by the spin-coating method, a promising approach for large area display applications as well as polymeric components.

Efforts have been ongoing to develop a novel hole-transport polymer for advanced electronic devices. This is a key material for improving the turn-on voltage, luminescence intensity, operation lifetime, full color display capability, durability, reasonable power efficiency, and so on. Generally, such high-performance devices are obtained by developing sequential HOMO/LUMO energy gradients in the device by introducing the multi-layered structure fabricated by repeatedly making thin films. For example, introducing another layer on the ITO electrode, with a HOMO energy level between that of the hole-transporting layer and of the electrode, that is a hole-injecting layer, lowers the energy barrier for hole injection. This approach involves placing the injecting layer between the ITO and the transport layers, such as copper phtharocyanine (CuPc),[9] and 4, 4', 4''-tris(3-methylphenylphenylamino)triphenylamine (m-MTDATA),[8] which results in an enhanced EL efficiency. Another approach to develop the characteristics of the OLEDs has recently been raised by the insertion of thick doped materials such as doped triarylamine,[10] polyvinylcarbazole,[11] polythiophene,[12] and polyaniline,[13] resulting in the enhancement of a carrier injection and transport with a lower driving voltage. However, these two approaches require highly-layered structures, because of their molecular structures and syntheses, and the indefinite functional separation of the buffer and hole transport in one component. Moreover, the use of polymeric materials is restricted by the fabrication of such very complex structures because of the erosion of the fabricated film in advance followed by the method of making thin films, i.e., the spin coating method.

Here, we report the design of novel carbazole dendrimers having a phenylazomethine core as the metal coordination site. Also, the complexation with metal ions changes the properties of the dendrimer. Only complexation with metal ions results in a higher EL efficiency.

Results and Discussion

Synthesis

The G1, G2, and G3 carbazole dendrimers[14] were synthesized via the cyclotrimerization of the corresponding monomers in the presence of titanium (IV) tetrachloride (0.75 equiv.) and 1,4-diazabicyclo-[2.2.2]octane (DABCO) (1.5 equiv.) in 30, 56, and 95% yields, respectively (Scheme 1). The aminophenylketone monomers were prepared by a previously described method.[16] As a conventional synthetic technique to prepare the dendrimers, the convergent method results in a lower conversion yield by increasing the generation due to their steric hindrance of the bulky dendrons with a small core.[17] We developed a useful synthetic method for preparative dendrimer formation in which the core is generated from a dendritic precursor by a cyclization reaction.[18] Generally, the condensation of the *AB* type monomer only gives the corresponding linear polymer. Indeed, we have already reported that aminophenylketones such as 4-aminoacetophenone produce only the corresponding linear polymer, however, the reaction of the 4-aminobenzophenone derivatives provides the formation of cyclic phenylazomethines in high yield due to the steric hindrance of the bulky α-phenyl ring in the monomer.[19] Thus, this synthetic procedure to prepare dendrimers, especially with a higher generation, via the cyclization reaction were found to be an extremely effective method. These dendrimers were characterized by ^{1}H and ^{13}C NMR spectroscopies, and matrix-assisted laser desorption/ionization time-of-flight mass spectroscopy (MALDI-TOF-MS).

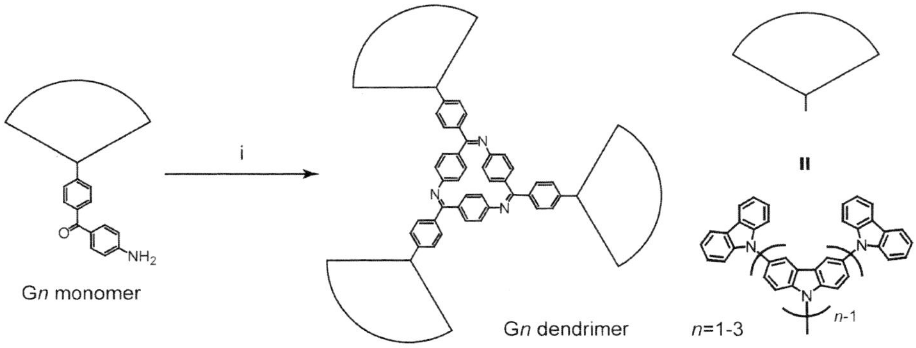

Scheme 1. Synthesis of the carbazole dendrimers[a]

[a] Reagents and conditions: (i) TiCl$_4$, DABCO, PhCl.

Optical Properties

The UV-vis spectra of the dendrimers show absorption bands at 300-425 nm (λ_{max}; 343 nm) attributed to the π-π* transition of the cyclic phenylazomethine unit at the core and at 225-350 nm (λ_{max}; 294 nm) of the dendritic polycarbazole moiety. No significant red shifts in the absorption were observed (Table 1). This means that the degree of π delocalization is limited.[27] The absorption attributed to the carbazole unit proportionally intensifies as the number of carbazole units with the growth of their generation number, even though the absorption attributed to their core unit does not change. This localization is caused by the loss of coplanarity of the aromatic systems due to the twists in the polycarbazole dendrons at the 3 and 6 positions. The MM2 calculation also supported the fact that they are twisted by 36 degrees between the carbazole units. These dendrimers are expected to approach the sphere-like structure with higher generations. The photoluminescence spectra showed luminescence from the carbazole moieties that were quenched by the Förster type energy transfer to the imine unit.

Metal-Collecting Properties

The addition of $SnCl_2$ to a dichloromethane/acetonitrile solution of the dendrimers resulted in a complexation with a spectral change, similar to that for the previously reported dendrimers.[20] During the addition of $SnCl_2$, the solution color of the G3 dendrimer changed from light to deeper yellow. We observed that the complexation was complete in 10 minutes by the spectral change after the addition of $SnCl_2$, that is, the complexation equilibrium is reached within at least several minutes.[21] Using UV-vis spectroscopy to profile the complexation, only one isosbestic point was observed, indicating that the complexation randomly proceeds. Therefore, this observation suggests that the complexation behavior seen with the dendrimers does not reflect changes in the basicity of the imine sites induced by complexation at neighboring imine sites.

The absorption band around 400 nm attributed to the complex increases with a decrease in the absorption bands around 320 nm, attributed to the phenylazomethine unit. The spectra of the G3 dendrimer gradually changed, with an isosbestic point at 371 nm (Figure 1a). However, the spectral change did not finish upon the addition of $SnCl_2$ equivalent to the number of imine moieties (three imines) not likely observed in the complexation of other phenylazomethine dendrimers with $SnCl_2$.[20] This observation is based on the poor electron density on the imine site supported by the chemical shifts of the imine carbon of the dendrimers (about 169.5 ppm) compared to those of the phenylazomethines (about 168 ppm). The dendrimers were decomposed with an increase in the free $SnCl_2$, leading to another type of spectral change. Complexing with $Eu(OTf)_3$ (OTf; trifluoromethane sulfonate) instead of $SnCl_2$ as a Lewis acid, the absorption band around 400 nm attributed to the complex increases with a decrease in the absorption bands around 320 nm, also attributed to the phenylazomethine unit in the same way. The spectra of the G3 dendrimer gradually changed with an isosbestic point at 381 nm (Figure 1b). The spectral change finished upon the addition of 10 equivalents of $SnCl_2$ to the dendrimer without any decomposition. These results revealed that the complexation proceeds at random and the imine groups act as an excellent coordination site.

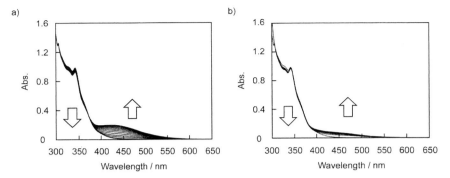

Figure 1. UV-vis spectra changes of the G3 dendrimer upon stepwise addition of equimolar a) SnCl$_2$ and b) Eu(OTf)$_3$ in CH$_3$CN/CH$_2$Cl$_2$ = 1/4.

Generally, a SnCl$_2$ molecule, which has one coordination site, is complexed with an imine site of phenylazomethine at the ratio of 1:1 (Figure 2a). On the other hand, for the complexation with Eu(OTf)$_3$, which has multi-coordination sites, the spectrum changed largely by addition of a small amount of Eu(OTf)$_3$ to the solution of the G3 dendrimer, resulting in a convex titration curve (Figure 2b). This phenomenon shows that a Eu(OTf)$_3$ molecule is complexed with several imine sites of the G3 dendrimer. The spectral change continued up to the addition of 10 equivalents of Eu(OTf)$_3$. This result shows that the complexation randomly occurs and the Eu salt is finally complexed with the imine site at the ratio of 1:1.

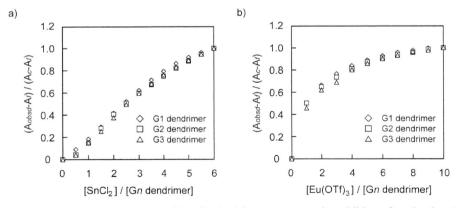

Figure 2. Titration curve changes of the G3 dendrimer upon stepwise addition of equimolar a) SnCl$_2$ and b) Eu(OTf)$_3$.

Comparing the generation number of the carbazole dendrimer, no significant difference in the complexation was observed, suggesting a dendrimer with an interior and an exterior with a small density up to the third generation. The equilibrium constants for the formation of 1:1 complexes with $SnCl_2$ (up to 6 equivalents to the dendrimer) and $Eu(OTf)_3$ on the imine sites were estimated to be $K = 10^5$ (M^{-1}). This estimation is based on the complexation behavior with the dendrimers indicating that the complexation randomly proceeds. We can transpose the cyclic phenylazomethine trimer unit to *three* monomeric phenylazomethine on the basis of the fact that the complexation at the neighboring imine site does not reflect changes in the basicity of the imine site.

Energy Levels of the Dendrimers

The electrochemical properties of the dendrimers (G1-G3) were studied by cyclic voltammetry. The HOMO values were determined from the first oxidation potential values with respect to ferrocene (-4.8 eV), as shown in Table 1. All the dendrimers displayed reversible oxidation waves attributed to the oxidation of triarylamines (the dendritic polycarbazole moiety) in the region between 0.2 V and 1.4 V *vs.* Fc/Fc$^+$. The oxidizability of the dendrimers decreased with the increasing generation number and decreasing electron richness, and therefore, the HOMO values gradually decreased from G3 (-5.2 eV) to G1 (-5.5 eV). Thus the most electron rich molecule, the G3 dendrimer, has the highest HOMO value of -5.2 eV and, accordingly, is expected to have the lowest barrier to hole injection from the ITO electrode (-4.8 eV) in OLEDs. By complexing with 1 equivalent of $Eu(OTf)_3$, redox waves with almost the same potential (-5.2 eV) as already observed in the oxidation wave of the dendron were seen. However, an additional stable redox wave attributed to the reduction of the azomethine-metal complex unit was not observed within the potential range.

Table 1. Physical data of the dendrimers

Compound	yield	λ_{max}	ε	$E_{1/2}^{ox}$	HOMO level
	%	nm	$mol^{-1}cm^{-1}$ [a)]	V vs.Fc/Fc$^+$ [b)]	eV
G1 dendrimer	30	292	5 700	0.647, 0.919	-5.5
		343	7 200		
G2 dendrimer	56	294	6 300	0.521, 0.795	-5.3
		342	9 800		
G3 dendrimer	95	294	12 100	0.383, 0.677, 0.883 [c)]	-5.2
		343	13 200		
G3-Eu complex				0.414, 0.676, 0.869 [c)]	-5.2

a) Measured in THF. b) Measured in 1,2-dichloroethane. c) Taken from differential pulse voltammetry peaks.

Electroluminescence Properties Using the Dendrimers as HTL

A bright green emission was observed for all the cells when a positive dc voltage was applied to the ITO electrode. The electroluminescence spectrum was in accord with the photoluminescence spectrum of Alq. This indicates that the electroluminescence originates from the recombination of holes and electrons injected into Alq. Compared with the generations of the dendrimers, the maximum luminescence was 67.8 (G1), 497.3 (G2), and 786.4 (G3) cd/m^2 (Figure 3, Table 2). Also, the *I-V* characteristics are strongly dependent on their generation number. For example, the current density at 10 V increased from 5.11 mA/cm^2 (G1) to 53.6 mA/cm^2 (G3). This development is explained by the barriers ($\Delta\phi$) between the HOMO energy levels of the dendrimers and that of ITO, which are lowered in higher generations, resulting in enhanced hole injection and charge recombination. However, the G1 dendrimer, which has the HOMO energy level at -5.5 eV, close to that of TPD (N,N'-diphenyl-N,N'-di(*m*-tolyl)benzidine) (-5.5 eV), which is a commonly used hole transport compound, has shown a lower EL performance with a poor film-forming property based on the planarity in the smaller generation number. A simple molecular modeling has revealed that the G1 dendrimer has a planar structure, thus resulting in the crystalline formation of the film.

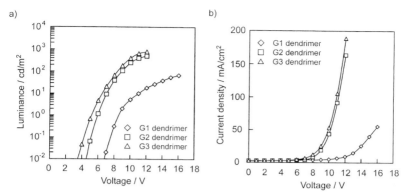

Figure 3. a) *L-V* and b) *I-V* characteristics of double-layer OLEDs with Al cathodes using the dendrimers as HTL.

For the two-layer OLEDs, the dendrimer complexes with the $Eu(OTf)_3$ films were then employed as the hole-transporting layer. We adopted $Eu(OTf)_3$ as a Lewis acid because of the stability of the dendrimer complex. From the spectral observation, the dendrimers were decomposed with an increase in the free $SnCl_2$. As shown in Figure 4 and Table 2, the low driving voltage and enhanced efficiency were followed by simply assembling of the metal ions. For the G3 dendrimer, the turn-on voltage was reduced from 4.3 V to 3.7 V and the maximum luminescence was enhanced from 786.4 cd/m^2 to 932 cd/m^2 by only complexing with $Eu(OTf)_3$. Also, the current performance of the G3-Eu complex was over three times greater (181.4 mA/cm^2) for the same forward driving voltage (10 V) compared to the uncomplexed device (53.6 mA/cm^2). The threshold voltages for obtaining a luminance of 100 cd/m^2 were 8.3 and 6.9 V for the cells, respectively. Considering the dendrimer complex as HTL, the characteristics of the devices are also strongly dependent on their generation number as seen in the device with the dendrimer itself as HTL. By inserting a thin CsF layer between the anode and electron transport layer, Alq, a significant improvement in the device has been obtained based on the enhancement of the electron injection and the performance of the device. The characteristics of the devices are also shown in Table 2. As measured, for the G3 dendrimer, the maximum luminescence was enhanced from 3 717 cd/m^2 to 6 718 cd/m^2 by applying a lower voltage, by only complexing with Eu(OTf)$_3$ under the non-optimized conditions. These results indicate that the complexation with imine sites results in an increasing hole injection and/or transport efficiency from the electrode.

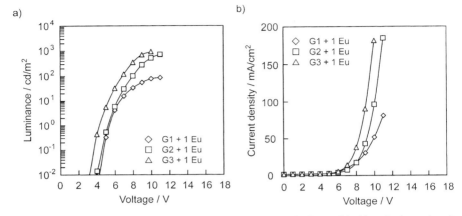

Figure 4. a) *L-V* and b) *I-V* characteristics of double-layer OLEDs with Al cathodes using the dendrimer complexes as HTL.

Table 2. Electroluminescence properties for the devices

HTLs	Turn-on voltage	Maximum luminance	Driving voltage	Current density	Power eff.
	V	cd/m^2	V [a]	mA/cm^{2} [a]	lm/W [a]
G1 dendrimer	7.6	67.8			
G2 dendrimer	5.1	497.3	8.9	17.6	0.20
G3 dendrimer	4.3	786.4	8.3	15.6	0.24
G1+Eu complex	5.0	86.7			
G2+Eu complex	4.0	728.1	8.0	17.0	0.23
G3+Eu complex	3.7	932.0	6.9	12.2	0.37
G3 dendrimer [b]	5.0	3717	8.2	7.66	0.50
G3+Eu complex [b]	3.6	6178	5.1	5.86	1.05

a) Taken at 100 cd/m^2. b) Measured with the double-layer OLEDs having CsF(3 nm)/Al cathodes.

We described the controlled doping levels of the dendrimer and the OLED characteristics. Figure 5 shows the I-V characteristics of the ITO/G3 dendrimer + n Eu/Alq/Al devices (n = 0, 1, 2, 3, 3.6, 6) with the same configuration. The devices with the dendrimer complexes having up to the small amount of metal ions with a small excess showed a significant injection current density hardly depending on the amount of metal ions. On the other hand, with a large excess of metal ions (6 equivalents to the dendrimer), a lower current density was injected into the device, resulting in the lower efficiency. This is due to the increased free Eu(OTf)$_3$ and its high crystallinity in the polymer film.

For the HOMO energy levels of the carbazole dendrimer complex, the energy levels of the carbazoles did not change based on the same redox potentials when complexing. When considering the influence of doping on the turn-on voltage and the performance at a lower driving voltage, the reduction in the bulk resistance followed by the p-type doping facilitates efficient hole injection into the HTL. Also the reduction of space charge layers at the interface of the ITO and the dendrimer complex leads to efficient carrier injection due to tunneling. Recently, a controlled doping study has revealed that p-type doping leads to a higher conductivity for the doped triphenylamine layers and a high density of equilibrium charge carriers, which facilitates hole injection and transport, and producing the low operating voltage of the OLEDs.[22] The conductivity of the G3 dendrimer film with metal ions increases and is orders of magnitude higher than those of the dendrimer itself. The room temperature conductivity, measured by a two-probe method, of the G3 dendrimer complex with 1 equivalent of Eu(OTf)$_3$ was 1.2×10^{-4} S/cm. For comparison, the conductivity of the G3 dendrimer is below 10^{-6} S/cm in an environmentally controlled condition. By complexation, higher hole injection and transport were facilitated, thus resulting in a lower operating voltage of the OLEDs.[20,23]

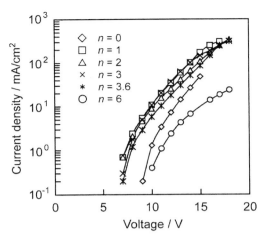

Figure 5. *I-V* characteristics of the ITO/G3 dendrimer + *n* Eu/Alq/Al devices (*n* = 0, 1, 2, 3, 3.6, 6).

Conclusion

A novel G3 carbazole dendrimer and its complex synthesized via the cyclotrimerization with a metal assembling property provide a homogeneous film which enhanced the performance of the electroluminescence device in comparison with only the G3 dendrimer as the hole-transporting layer. This is due to the enhancement of the conductivity by the *p*-type doping of the imine sites, resulting in better hole injection and transporting properties. This approach was managed by a solution process, i.e., spin-coating method, without using the coevaporation technique based on the large equilibrium constants of the coordination of metal ions on the imine sites (K = 10^5 M^{-1}).

Experimental Section

All chemicals were purchased from Aldrich, Tokyo Kasei Co., Ltd., and Kantoh Kagaku Co., Inc. (reagent grade) and used without further purification. The UV-vis spectra were obtained using a Shimadzu UV-2400PC spectrometer. The ^{1}H NMR and ^{13}C NMR spectra were measured on a JEOL 400 MHz FT-NMR (JMN 400). Cyclic voltammetry experiments were performed with a BAS-100 electrochemistry analyzer. All measurements were carried out at room temperature with a conventional three-electrode configuration consisting of a platinum working electrode, an auxiliary platinum electrode, and a nonaqueous $Ag/AgNO_3$ reference electrode. The solvent in all experiments was 1,2-dichloroethane, and the supporting electrolyte was 0.1 M tetrabutylammonium hexafluorophosphate. The $E_{1/2}$ values were determined as $1/2(E_{pa}+E_{pc})$, where E_{pa} and E_{pc} are the anodic and cathodic peak potentials, respectively. All reported potentials are referenced to Fc^+/Fc external and are not corrected for the junction potential.

Conventional OLED devices having the ITO/dendrimer/Alq/Al structure were fabricated by spin-coating the dendrimer solutions in chlorobenzene on an ITO-coated glass anode. Alq (50 nm), and Al (100 nm) were successively vacuum deposited on top of the hole-transporting layer. The emitting area was 9 mm^2. The current-voltage characteristics were measured using an Advantest R6243 current/voltage unit. Luminance was measured with a Minolta LS-100 luminance meter in air at room temperature.

For the two-layer OLEDs, the dendrimer complexes with the $Eu(OTf)_3$ films, were then employed as the hole-transporting layer. To a solution of the dendrimer in chloroform was added a solution of $Eu(OTf)_3$ (1 equivalent $vs.$ the dendrimers) in acetonitrile. The yellow solution changed to light orange based on the complexation, and evaporated to dryness to give the dendrimer complex. OLED devices having the ITO/dendrimer/Alq/CsF/Al structure were fabricated by spin-coating the dendrimer complex solutions in chlorobenzene on an ITO-coated glass anode. Alq (50 nm), CsF (2 nm) and Al (100 nm) were then successively vacuum deposited on top of the hole-transporting layer.

General Procedure for the Synthesis of the Gn Carbazole Dendrimers

To a mixture of the Gn monomer (n=1-3) and DABCO in chlorobenzene was added dropwise $TiCl_4$. The addition funnel was then rinsed with chlorobenzene. The reaction mixture was heated in an oil bath at 125 °C for 22 h. The precipitate was removed by filtration. The filtrate was concentrated, and the residue was subjected to recycling SEC with THF as the eluent, where the fraction corresponding to the second major peak was collected and evaporated to dryness, to give the corresponding dendrimer

G1 Dendrimer. The previous procedure was followed using 0.272 g (0.187 mmol) of the G1 monomer, 0.246 g (2.19 mmol) of DABCO, and 0.104 g (0.547 mmol) of $TiCl_4$ in chlorobenzene (7.5 mL), to produce the G1 dendrimer as a yellow powder in 30% yield (0.077 g). ^1H NMR

(400 MHz, CDCl$_3$, TMS standard, 30 °C, ppm): δ 8.16 (d, 7.6 Hz, 6H), 8.15 (d, J=8.2 Hz, 6H), 7.69 (d, J=7.6 Hz, 6H), 7.52 (d, J=7.6 Hz, 6H), 7.44 (dd, J=7.4, 8.2 Hz, 6H), 7.32 (dd, J=7.6, 7.4 Hz, 6H), 6.95 (d, 7.6 Hz, 6H), 6.66 (d, J=7.6 Hz, 6H). ^{13}C NMR (100 MHz, CDCl$_3$, TMS standard, 30 °C, ppm): δ 170.23, 152.63, 140.35, 140.27, 136.62, 130.61, 128.10, 126.38, 126.00, 123.57, 120.30, 120.25, 119.73, 109.72. MALDI-TOF-MS: 1032.4 ([M]$^+$ calcd for C$_{75}$H$_{48}$N$_6$: 1032.4).

G2 Dendrimer. The previous procedure was followed using 0.520 g (0.750 mmol) of the G2 monomer, 0.253 g (2.25 mmol) of DABCO, and 0.107 g (0.563 mmol) of TiCl$_4$ in chlorobenzene (7.5 mL), to produce the G2 dendrimer as a yellow powder in 56% yield (0.283 g). ^{1}H NMR (400 MHz, CDCl$_3$, TMS standard, 30 °C, ppm): δ 8.30(s, 6H), 8.28(d, J=11.2 Hz, 6H), 8.17(d, J=7.2 Hz, 12H), 7.87(d, J=8.8 Hz, 6H), 7.76(d, J=8.8 Hz, 6H), 7.65(d, J=8.8 Hz, 6H), 7.41(s, 24H), 7.31-7.25(m, 12H), 7.02(d, J=8.0 Hz, 6H), 6.73(d, J=8.0 Hz, 6H). ^{13}C NMR (100 MHz, CDCl$_3$, TMS standard, 30 °C, ppm): δ 170.23, 152.71, 141.59, 140.13, 139.78, 137.43, 137.43, 135.64, 131.00, 130.69, 128.21, 126.65, 126.35, 125.85, 125.44, 124.23, 123.10, 120.27, 119.78, 119.69, 111.23, 109.59. MALDI-TOF-MS: 2024.3 ([M + H]$^+$ calcd for C$_{147}$H$_{90}$N$_{12}$: 2022.7).

G3 Dendrimer. The previous procedure was followed using 0.494 g (0.365 mmol) of the G3 monomer, 0.123 g (1.09 mmol) of DABCO, and 0.052 g (0.274 mmol) of TiCl$_4$ in chlorobenzene (5.5 mL), to produce the G3 dendrimer as a yellow powder in 95% yield (0.459 g). ^{1}H NMR (400 MHz, CDCl$_3$, TMS standard, 30 °C, ppm): δ 8.52(m, 168H), 6.75(m, 6H). ^{13}C NMR (100 MHz, CDCl$_3$, TMS standard, 30 °C, ppm): δ 170.15, 152.72, 141.64, 141.40, 140.63, 140.35, 139.47, 137.74, 135.60, 131.16, 130.18, 130.01, 129.46, 128.13, 126.46, 126.22, 125.78, 125.43, 124.34, 123.69, 123.60, 123.01, 120.21, 119.60, 111.71, 111.13, 109.57. MALDI-TOF-MS: 4009.6 ([M + H]$^+$ calcd for C$_{291}$H$_{174}$N$_{24}$: 4003.4).

Acknowledgement

This work was partially supported by CREST from Japan Science and Technology Agency, Grants-in-Aid for Scientific Research (Nos. 15036262, 15655019、15350073) and the 21st COE Program (Keio-LCC) from the Ministry of Education, Science, Culture and Sports, and a Research Grant(Project No. 23) from Kanagawa Academy Science and Technology.

[1]   F. E. Goodson, S. I. Hauck, J. F. Hartwig, *J. Am. Chem. Soc.* **1999**, *121*, 7527.

[2]   T. D. Selby, S. C. Blackstock, *Org. Lett.* **1999**, *1*, 2053.

[3]   [3a] J. Louie, J. F. Hartwig, *J. Am. Chem. Soc.* **1997**, 119, 11695; [3b] T. D. Selby, S. C. Blackstock, *J. Am. Chem. Soc.* **1998**, *120*, 12155; [3c] M. I. Ranasinghe, O. P. Varnavski, J. Pawlas, S. I. Hauck, J. Louie, J. F. Hartwig, T. Goodson III, *J. Am. Chem. Soc.* **2002**, *124*, 6520.

[4]   C. W. Tang, S. A. VanSlyke, *Appl. Phys. Lett.* **1987**, *51*, 913.

[5]   U. Bach, D. Lupo, P. Conte, J. E. Moser, F. Weissörtel, J. Salbeck, H. Spreitzer, M. Grätzel, *Nature,* **1998**, *395*, 583.

[6]   M. Higuchi, S. Shiki, K. Ariga, K. Yamamoto, *J. Am. Chem. Soc.* **2001**, *123*, 4414.

[7]   [7a] J. Hofkens, M. Maus, T. Gensch, T. Vosch, M. Cotlet, F. Köhn, A. Herrmann, K. Müllen, F. De Schryver, *J. Am. Chem. Soc.* **2000**, *122*, 9278; [7b] M. Schlupp, T. Weil, A. J. Berresheim, U. M. Wiesler, J. Bargon, K. Müllen, *Angew Chem. Int. Ed.,* **2001**, *40*, 4011.

[8]   [8a] Y. Shirota, Y. Kuwabara, H. Inada, T. Wakimoto, H. Nakada, Y. Yonemoto, S. Kawami, K. Imai, *Appl. Phys. Lett.* **1994**, *65*, 807; [8b] P.-W. Wang, Y.-J. Lin, C. Devadoss, P. Bharathi, J. S. Moore, *Adv. Mater.* **1996**, *8*, 237; [8c] A. W. Freeman, S. C. Koene, P. R. L. Malenfant, M. E. Thompson, J. M. J. Fréchet, *J. Am. Chem. Soc.* **2000**, *122*, 12385; [8d] S. C. Lo, N. A. H. Male, J. P. J. Markham, S. W. Magennis, P. L. Burn, O. V. Salata, I. D. W. Samuel, *Adv. Mater.* **2002**, *14*, 975; [8e] P. Furuta, J. Brooks, M. E. Thompson, J. M. J. Fréchet, *J. Am. Chem. Soc.* **2003**, *125*, 13165.

[9]   S. A. VanSlyke, C. H. Chen, C. W. Tang, *Appl. Phys. Lett.* **1996**, *69*, 2160.

[10]   [10a] F. Huang, A. G. MacDiamid, B. R. Hsieh, *Appl. Phys. Lett.* **1995**, 67, 1659; [10b] X. Zhou, M. Pfeiffer, J. Blochwitz, A. Werner, A. Nollau, T. Fritz, K. Leo, *Appl. Phys. Lett.* **2001**, *78*, 410.

[11]   R. H. Partridge, *Polymer* **1997**, *81*, 3294.

[12]   [12a] S. Hayashi, H. Etoh, S. Saito, *Jpn. J. Appl. Phys., Part 2* **1986**, *25*, L773; [12b] D. B. Romero, M. Schaer, L. Zuppiroli, B. Cesar, B. Francois, *Appl. Phys. Lett.* **1995**, *67*, 1659; [12c] S. A. Carter, M. Angelopoulos, S. Karg, P. J. Brock, J. C. Scott, *Appl. Phys. Lett.* **1997**, *70*, 2067; [12d] M. Gross, D. C. Müller, H-G. Nothofer, U. Scherf, D. Neher, C. Bräuchle, K. Meerholz, *Nature* **2000**, *405*, 661.

[13]   [13a] Y. Yang, A. J. Heeger, *Appl. Phys. Lett.* **1994**, *64*, 1245; [13b] J. R. Sheets, H. Antoniadis, M. R. Hueschen, W. Lepnard, J. Miller, R. Moon, D. Roitman, A. Stocking, *Science* **1996**, *273*, 884.

[14]   Carbazole dendrimers with a conjugated backbone have been the subject of recent studies.[15]

[15]   [15a] Z. Zhu, J. S. Moore, *J. Org. Chem.* **2000**, *65*, 116; [15b] N. D. McClenaghan, R. Passalacqua, F. Loiseau, S. Campagna, B. Verheyde, A. Hameurlaine, W. Dehaen, *J. Am. Chem. Soc.* **2003**, *125*, 5356.

[16]   A. Kimoto, J-S. Cho, M. Higuchi, K. Yamamoto, *Chem. Lett.* **2003**, *32*, 674.

[17]   C. J. Hawker, J. M. J. Fréchet, *J. Am. Chem. Soc.* **1990**, *112*, 7638.

[18]   M. Higuchi, H. Kanazawa, M. Tsuruta, K. Yamamoto, *Macromolecules* **2001**, *34*, 8847.

[19]   M. Higuchi, A. Kimoto, S. Shiki, K. Yamamoto, *J. Org. Chem.* **2000**, *65*, 5680.

[20]   [20a] K. Yamamoto, M. Higuchi, S. Shiki, M. Tsuruta, H. Chiba, *Nature* **2002**, 415, 509; [20b] T. Imaoka, H. Horiguchi, K. Yamamoto, *J. Am. Chem. Soc.* **2003**, *125*, 340; [20c] N. Sato, J-S. Cho, M. Higuchi, K. Yamamoto, *J. Am. Chem. Soc.* **2003**, 125, 8104; [20d] M. Higuchi, M. Tsuruta, H. Chiba, S. Shiki, K. Yamamoto, *J. Am. Chem. Soc.* **2003**, *125*, 9988.

[21]   The absorption band around 600 nm attributed to the complexation of $SnCl_2$ with carbazole unit has not been observed during the complexation with imine units because of the larger equilibrium constants of the imine units.

[22]   [22a] M. Pfeiffer, A. Beyer, T. Fritz, K. Leo, *Appl. Phys. Lett.* **1998**, *73*, 3202; [22b] X. Zhou, J. Blochwitz, M. Pfeiffer, A. Nollau, T. Fritz, K. Leo, *Adv. Funct. Mater.* **2001**, *11*, 310; [22c] A. Yamamori, C. Adachi, T. Koyama, Y. Taniguchi, *J. Appl. Phys.* **1999**, *86*, 4369.69.

[23]   [23a] A. Kimoto, J-S. Cho, K. Yamamoto, *J. Photopolym. Sci. Technol.* **2003**, *16*, 293; [23b] J-S. Cho, A. Kimoto, T. Nishiumi, K. Yamamoto, *J. Photopolym. Sci. Technol.* **2003**, *16*, 295.

Macromol. Symp. **2004**, *209*, 67-79

Molecular Weight Determination of Coordination and Organometallic Oligomers by T_1 and NOE Constant Measurements : Concepts and Challenges

Pierre D. Harvey

Département de chimie, Université de Sherbrooke, Sherbrooke, PQ, Canada, J1K 2R1

E-mail: pierre.harvey@usherbrooke.ca

Summary: This paper describes how to determine molecular weights of coordination and organometallic polymers (or rather oligomers) in solution using spin-lattice relaxation time (T_1) and Nuclear Overhauser Enhancement constant (η_{NOE}) measurements. The methodology is explained using simple organometallic complexes such as $M(CN\text{-}t\text{-}Bu)_4^+$ complexes (M = Cu, Ag). Very good results are obtained for oligomers that exhibit a rigid structure. Conversely, very poor results are extracted when the materials show flexible chains in the backbone. The typical examples for rigid and flexible oligomers are the $\{Ag(dmb)_2^+\}_n$ (dmb = 1,8-diisocyano-p-menthane), and $\{Pd_2(dmb)_2(diphos)^{2+}\}_n$ (diphos = dppa, dppb, dpppent, and dpph) as well as $\{Pd_2(diphos)_2(dmb)^{2+}\}_n$ (diphos = dppe, dppr, and dppp**R**; **R** = $O(CH_2)_2O$-naphthyl), respectively.

Keywords: molecular dynamics; molecular weight measurements; Nuclear Overhauser Enhancement; spin-lattice relaxation times

Introduction

The molecular weight (M_n or M_w) and molecular weight distribution (also called the poly-dispersity constant) are two of the fundamental properties of macromolecules. In the area of coordination and organometallic polymers, a particularly challenging problem occurs when the polymeric materials turn out to be oligomeric in solution. The problem is the accurate estimation of smaller molecular dimension polymers (i.e. M_n or M_w). Techniques such as light scattering, osmometric and intrinsic viscosity measurements are three methods commonly used to determine these properties. However, a limitation of 5 000-10 000 exists for these methods, making the investigations of oligomers difficult. In addition, the relatively weak solubility that some materials may exhibit represents another important challenge for the techniques mentioned above.

© 2004 WILEY-VCH Verlag GmbH & KGaA, Weinheim

DOI: 10.1002/masy.200450505

On the other hand, techniques such as Fast Atom Bombardments (FAB), Matrix-Assisted-Laser-Desorption-Ionisation-Time-of-Flight (MALDI-TOF), and electron spray mass spectrometry are also common characterization techniques for materials with molecular weight exceeding 1 000-1 300 mass units. Because of the relative fragility of the coordination bond or the fact that some species are already charged like in poly-cationic polymers, there is no guarantee of success in such measurements.

Because of all these common problems encountered by our group, another methodology that overcomes solubility and molecular dimension challenges has been used. This method is particularly suitable for oligomers, and is based on NMR techniques called spin-lattice relaxation time and Overhauser Nuclear Enhancement constant (η_{NOE}), which seeks to evaluate the dimension of the tumbling molecule from the knowledge of the correlation time (τ_c). This paper describes the concept and methodology, and some examples where this method does and does not apply.

Concept

According to theory, the volume (V) of a spherical molecule tumbling in solution can be calculated by using the Stokes-Einstein-Debye equation (1):

$$\tau_c = V \, \eta_{visc} \, / \, k \, T \qquad\qquad (1)$$

where η_{visc} is the viscosity of the medium, k is the Boltzman constant, and T is the temperature. For non-spherical molecules, this equation becomes more complex and involves a decomposition of the molecular motions according to the various axes of the molecule (normally x, y, and z, for ellipsoid-shaped molecules), and as a consequence, three correlation times must be obtained. Because the T_1 measurements provide an average value, it is inadequate to attempt to decompose the problem along the three axes. In such a case, this method is an approximation.

The measurement of the dipole-dipole relaxation time (T_1^{DD} = a spin relaxation due to interactions with the surrounding nuclei), one can extract τ_c according to the general Equation 2 (for this equation the considered dipole-dipole interactions are those generated by the ^1H and ^{13}C nuclei generally encountered in ^{13}C NMR for instance):

$$1/T_1^{DD} = \Sigma(\hbar^2\gamma_H^2\gamma_C^2/10r_{CH}^6)\bullet\{\tau_c/[1+(\omega_H-\omega_C)^2\tau_c^2]+3\tau_c/[1+\omega_C^2\tau_c^2] + 6\tau_c/[1+(\omega_H+\omega_C)^2\tau_c^2]\} \quad (2)$$

where $\hbar$ $(h/2\pi)$ is the Plank constant, γ is the magnetogyric ratio for a given nuclei, and r_{CH} is the distance between the two interacting nuclei. In the extreme narrowing limit, where molecular motions are very fast (i.e. small molecules, $\omega\tau_c \ll 1$), Equation 2 becomes:

$$1/T_1^{DD} = \Sigma [(\hbar^2 \gamma_H^2 \gamma_C^2) / (r_{CH}^6)] \bullet \tau_c \quad (3)$$

To verify if the extreme narrowing limit applies, one has to check if T_1 varies with the magnetic field strength (H_o). If it does not apply, then Equation 2 must be used. Experimentally, the measured T_1 represents the sum of all contributions to spin-lattice relaxations, so that T_1^{DD} must be extracted from:

$$1/T_1 = \Sigma 1/T_1^i = 1/T_1^{DD} + 1/T_1^{SR} + 1/T_1^{CSA} + 1/T_1^{SC} + 1/T_1^{exc} + 1/T_1^Q + 1/T_1^{para} \quad (4)$$

where SR, SCA, SC, exc, and para designate relaxations operating via spin-rotation, chemical shift anisotropy, scalar, chemical exchange, quadrupolar, and the presence of paramagnetic electrons (details for the description of these different mechanisms to relaxation are given in ref.[1] and ref.[2]. In the extreme narrowing limit, T_1^{DD} is field independent and can be extracted at the intercept of the graph $1/T_1$ vs H_o. For such a method, accuracy can only be obtained if many spectrometers with different fields are available, which is rarely the case. Instead, the measurements of the NOE effect (the experimentally observed enhancement of the ^{13}C signal, for instance, during coupled vs uncoupled experiments) give access to η_{NOE}:

$$NOE = 1 + \eta_{NOE} = 1 + (\gamma_H/2\gamma_C) \quad (5)$$

where η_{NOE} is the fractional NOE constant. Hence one can more confidently obtain T_1^{DD}:

$$1 / T_1^{DD} = (\eta_{NOE})/(T_1 \bullet \eta_{max}) \tag{6}$$

where η_{max} is the maximum NOE effect one can get for two given nuclei in the extreme narrowing limit. For ^1H and ^{13}C interactions, $\eta_{max} = \gamma_H/2\gamma_C = 1.988$, while for ^1H and ^{31}P, for example, $\eta_{max} = \gamma_H/2\gamma_P = 1.235$, where the ^{13}C and ^{31}P are the probed nuclei (i.e. for which the T_1 are measured).

For example, the tetracoordinated cations $M(CN\text{-}t\text{-}Bu)_4^+$ (M= Cu, Ag) have recently been investigated as models for the repetitive unit in the polycationic coordination polymer $\{Ag(dmb)_2^+\}_n$.[3, 4] Table 1 compares the T_1 data for the various salts.

Table 1. T_1 data for $[M(CN\text{-}t\text{-}Bu)_4](X)$ salts for the 63-64 ppm signal ($-^{13}C(CH_3)_4$)[a]

M	X	T_1
		s
Ag	BF_4^-	37.5
Ag	ClO_4^-	37.1
Ag	NO_3^-	36.0
Cu	BF_4^-	34.2
Cu	ClO_4^-	32.6
Cu	NO_3^-	33.5

[a] At 298 K in d_3-acetonitrile on a 300 MHz instrument.

There is very little influence of the counter-anion on T_1, indicating that the dragging of the counter-ion during tumbling could be neglected. Similarly, the close resemblance of $T_1(Cu)$ and $T_1(Ag)$ indicates that both nuclei have a similar influence on T_1 for this specific nuclei. The η_{NOE} data are also about constant ($\eta_{NOE} = 1.31 \pm 0.03$). For $[Ag(CN\text{-}t\text{-}Bu)_4]BF_4$ as an example, equation 6 computes $T_1^{DD} = 57.0$ s, and τ_c (from equation 3) is 2.11×10^{-10} s. The "hydrodynamic" volume that takes into account the dragging of solvent molecules (and counter-ions) during tumbling is 2.37×10^{-27} m^3 (or 2370 Å^3) according to the Stokes-Einstein-Debye equation. The use of this relation is justified by the fact that the tetrahedral structure of the cation resembles that of a sphere. The value for V is very large when compared to X-ray data of related complex $Fe(CN-t-Bu)_5$ (in the order of 750 Å^3)[5] (since no structure exists for the $[M(CN-t-Bu)_4](X)$ complexes discussed above). This difference not only shows the importance of the solvent dragging, but also the potentially large inaccuracy that one may experience.

In order to overcome this problem, one can extract the dimension of an unknown molecule using another molecule for comparison (i.e. a standard). This molecule must have some important characteristics to be adequate. The existence of an X-ray structure is preferable in order to obtain information about the volume. To the contrary, computer modeling is sufficient. The standard should also bear the same charge as the unknown, as well as the same ligands. These properties confirm that the dragging of the counter-ion and solvent molecules during tumbling in solution are the same. Hence, by combining Equation 1, 3 and 6, one obtains:

$$\frac{T_1(sam)}{T_1(sta)} = \frac{\eta_{CH}(sam)}{\eta_{CH}(sta)} \cdot \frac{V(sta)}{V(sam)} \cdot \frac{\Sigma \ 1/r^6_{CH}(sta)}{\Sigma \ 1/r^6_{CH}(sam)} \qquad (7)$$

where sta and sam stand for standard and sample, respectively, and η_{CH} is η_{NOE} (applied to ^{13}C-^1H interactions as an example). By using exactly the same probe on the same ligand in the standard and the sample, the term $[\Sigma \ 1/r^6_{CH}(sam)]/ \ [\Sigma \ 1/r^6_{CH}(sta)]$ becomes unity. Hence, to extract V(sam), one measures $T_1(sam)$, $T_1(sta)$, $\eta_{CH}(sam)$, and $\eta_{CH}(sta)$, and uses X-ray data or computations for V(sta), again if the extreme narrowing limit applies. For organometallic or coordination polymers, by using a standard that closely resembles a repetitive unit of the polymer, the ratio V(sam)/V(sta) can give a very good estimate of the number of units in the polymeric chain.

Two problems exist; 1) the use of the Stoke-Einstein-Debye equation is an approximation in this methodology, and 2) the standard does not always perfectly mimic the solute-solvent interactions as the polymer chain does, for simple geometry reasons. Despite these problems, the estimation of the molecular dimension is still valuable information, when no other technique can be applied.

The {Ag(dmb)$_2^+$}$_n$ Case

Since the discovery of this polymer in 1992 (Scheme 1)[6] and its Cu analogue,[7] many new related polymers such as {Pt$_4$(dmb)$_4$(diphos)$_2^+$}$_n$,[8] (diphos = Ph$_2$P(CH$_2$)$_m$PPh$_2$ (m = 4-6), {Pd$_4$(dmb)$_4$(dmb)$^{2+}$}$_n$,[9] mixed-metal {Cu$_x$Ag$_{1-x}$(dmb)$_2^+$}$_n$,[10] mixed-ligand {M(dmb)-(dppm)$^+$}$_n$ (M = Cu, Ag; dppm= Ph$_2$PCH$_2$PPh$_2$),[10] and mixed-conformation[11] have been prepared and confidently characterized from X-ray crystallography or, on rare occasions, molecular weight determination.

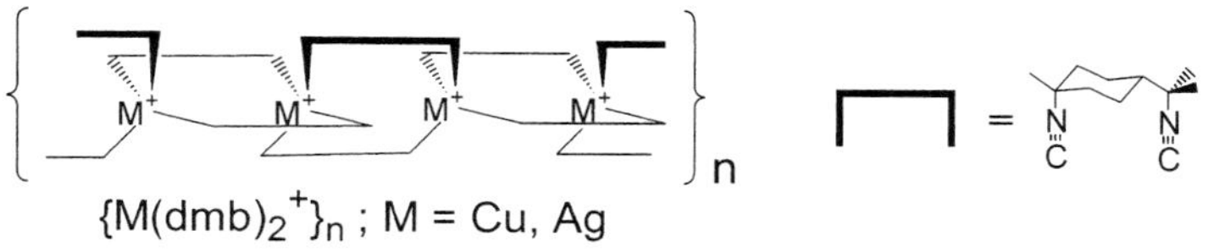

{M(dmb)$_2^+$}$_n$; M = Cu, Ag

Scheme 1

For the title polymer, light scattering and osmometric measurements indicate that M$_w$ and M$_n$ cannot be measured by these methods, which place the upper limit to 10 000. Similarly, both FAB mass and MALDI-TOF spectra exhibit fragment peaks of about 3 units (1100-1300).[5] The measurements of the intrinsic viscosity exhibit a delay of barely 0.5 s, which make this method far too inaccurate for any meaningful use.

For the {[Ag(dmb)$_2$]X}$_n$ polymers (X = BF$_4^-$, ClO$_4^-$, NO$_3^-$), the probe nuclei is the sp^3-hybridized quaternary carbon #2, more particularly ^{13}C located in the dmb-cyclohexyl ring (see Scheme 2). This nuclei is selected for the following reasons. The rigidity of the M-C≡N-^{13}C unit allows one to measure a T$_1$ process that is less influenced by intramolecular motions such as spin-rotation relaxation known for methyl groups for instance, and motion of alkyl chains. In addition a quaternary ^{13}C relaxes more slowly than others, because of the absence of close ^{1}H···^{13}C dipole-dipole interactions which are known to induce efficient relaxation of the ^{13}C nuclei. As a result the larger time scale provides an opportunity to gain accuracy. The δ value for this ^{13}C is 63.3 ppm (CD$_3$CN).

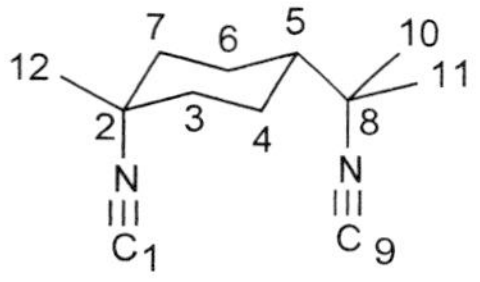

Scheme 2

Table 2 summarizes the NMR results for various salts of $\{[Ag(dmb)_2]X\}_n$ ($X = BF_4^-$, ClO_4^-, NO_3^-). The average η_{NOE} in these cases is 1.75 ± 0.03, expectedly different from that found for the corresponding ^{13}C nuclei in the $[Ag(CN-t-Bu)_4](X)$ salts used as standards due to the great similarity between two Bu-t-NC ligands and one dmb (Scheme 3).

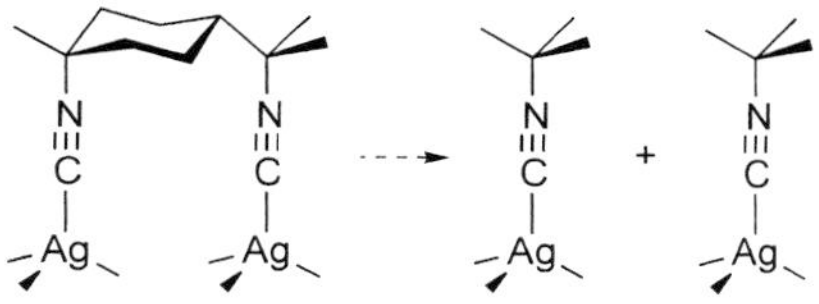

Scheme 3

The V(sta) is the hydrodynamic volume estimated above, and it is assumed that $V(Ag(CN-t-Bu)_4^+) \sim V(Ag(dmb)_2^+)$. The number of units ranges from 7 to 9 giving M_n of 4 000-5 000, which is exactly where it was predicted to be (see above). These oligomers correspond to small "rigid rods" of about 43-50 Å (or 4.5 – 5 nm) in lenght.

Table 2. Selected data for the $\{[Ag(dmb)_2]X\}_n$ materials.

	NO_3^-	BF_4^-	ClO_4^-
T_1 (± 0.20 s)	6.90	7.53	7.93
T_1^{DD}(± 0.20 s)	7.84	8.55	9.77
τ_c (x 10-9 s $\pm 3\%$)	1.97	1.81	1.58
V (Å^3)	22000	203000	17700
Number of units	~9	~8	~7

Larger Polymers

The corresponding $\{[Cu(dmb)_2]BF_4\}_n$ polymer, M_w of 160 000[7] and M_n of 133 000[12] (about 300 units) are obtained from light scattering and osmometric measurements, respectively. On the basis of T_1, one can predict that as T_1 decreases, the molecular dimension increases. This is well exemplified by the series $\{Ag(dmb)_2^+\}_n$ (T_1 = 6.90-7.93 s; number of units = 8 ± 1 above), $\{Ag_{0.27}Cu_{0.73}(dmb)_2^+\}_n$ (T_1 = 6.35 s),[4] and $\{Cu(dmb)_2^+\}_n$ (T_1 = 2.63 s; ~300 units). The change of T_1 vs the number of units is not linear, as one would anticipate from Equation 2, which in fact exhibits a V-shaped curve. As T_1 decreases, the method becomes more inaccurate due to the lack of sensitivity of T_1 vs M_n (or τ_c). This phenomena is also well demonstrated in the cases of the $\{Pt_4(dmb)_4(diphos)_2^+\}_n$ (diphos = $Ph_2P(CH_2)_mPPh_2$ (m = 4-6; Scheme 4, Table 3) polymers for which M_n was obtained from the intrinsic viscosity measurements,[8] reported earlier. The η_{NOE} values were not measured, so the comparison between the number of units measured from the intrinsic viscosity properties and from the T_1/NOE method could not be performed.

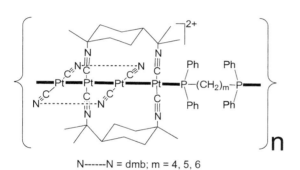

N------N = dmb; m = 4, 5, 6

Scheme 4

Table 3. Dimension of the $\{Pt_4(dmb)_4(Ph_2P(CH_2)_mPPh_2)\}_n$ polymers.[a]

Polymers	$[\eta]$	T_1	M_n	numbers
		^{13}C		of units
	cm^3/g	s		
$Pt_4(dmb)_4(PPh_3)_2{}^{2+}$	—	3.76	2000	1
$\{Pt_4(dmb)_4(Ph_2P(CH_2)_4PPh_2)^{2+}\}_n$	3.66	2.19	203000	100
$\{Pt_4(dmb)_4(Ph_2P(CH_2)_5PPh_2)^{2+}\}_n$	4.78	1.69	307000	150
$\{Pt_4(dmb)_4(Ph_2P(CH_2)_6PPh_2)^{2+}\}_n$	2.06	2.24	84000	40

[a]The same dmb-peak at 63 ppm is probed. The solvent is CD_3CN at room temperature.

In conclusion, this methodology is better suited for small oligomers, which is very convenient here, since reliable methods for small molecules and larger polymers already exist.

Molecular Dynamics

As stated above, the comparison between the number of units obtained from the intrinsic viscosity properties and from the T_1/NOE method could not be performed for this earlier work, but was recently performed in our laboratory[13] Indeed, two series of organometallic polymers have been prepared (Scheme 5): 1) $\{Pd_2(diphos)_2(dmb)^{2+}\}_n$ (diphos = dppe, dpppr, dpppR (R = O-$(CH_2)_6$-O-Naphthyl)), and 2) $\{Pd_2(dmb)_2(diphos`)^{2+}\}_n$ (diphos = dppb, dpppent, dpph, dppa). Because the ligand environment about the Pd_2-bonded fragment is different, the color of the materials is different. These are respectively, yellow and red, and are found to be amorphous according to XRD patterns. Both series exhibit the dmb ligand, which can act as a bridging ligand between two Pd_2 units, or the Pd-Pd bond.

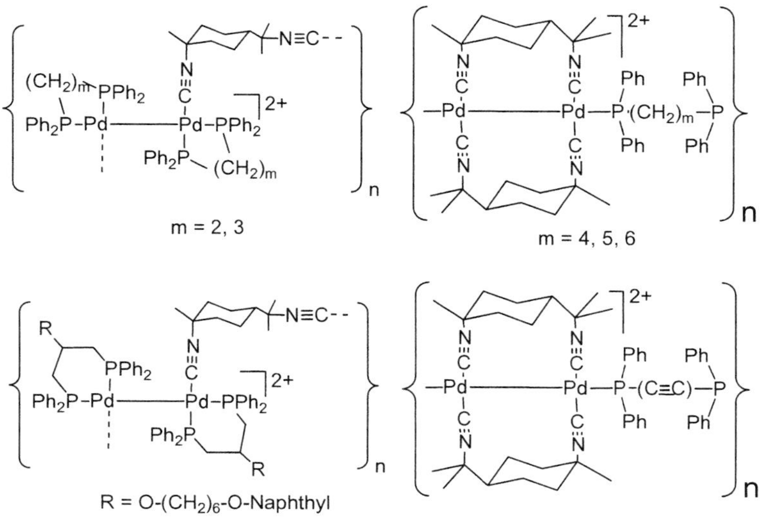

Scheme 5

The two T_1/NOE standards used for the two series are the complexes $Pd_2(dmb)_2(PPh_3)_2{}^{2+}$ and $Pd_2(dppe)_2(dmb)_2{}^{4+}$ (Scheme 6) for the "red" and "yellow" series, respectively.

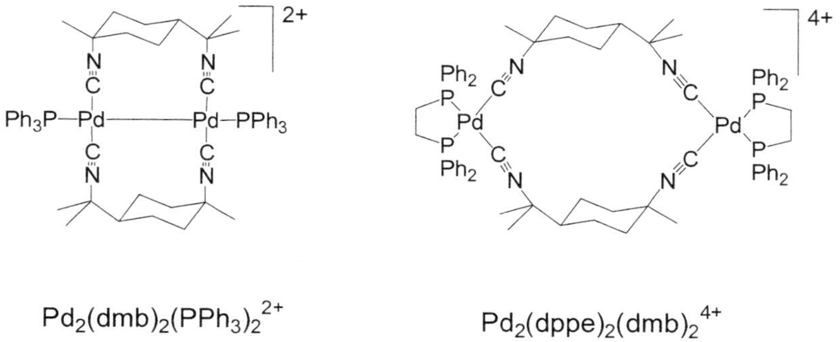

$Pd_2(dmb)_2(PPh_3)_2{}^{2+}$ $\qquad$ $Pd_2(dppe)_2(dmb)_2{}^{4+}$

Scheme 6

The data in Table 4 summarizes the M_n data obtained from the measurements of the intrinsic viscosity, and the corresponding number of units. The "yellow" series, $\{Pd_2(diphos)_2\text{-}(dmb)^{2+}\}_n$ (diphos = dppe, dpppr, dpppR), are found to be a little shorter (~7-8) than that of the "red" series, $\{Pd_2(dmb)_2(diphos`)^{2+}\}_n$ (diphos = dppb, dpppent, dpph, dppa) (12-16 units). These materials are oligomers when dissolved in solution (which does not necessarily reflect the nature of the materials in the solid state).

Table 4. Comparison between T_1/NOE and intrinsic viscosity data

	M_n	Nb. of units (intr. visc.)	Nb. of units (T_1/NOE)
$\{Pd_2(dmb)_2(dppb)^{2+}\}_n$	17 800	~14	~ 4
$\{Pd_2(dmb)_2(dpppen)^{2+}\}_n$	18 400	~14	~ 4
$\{Pd_2(dmb)_2(dpph)^{2+}\}_n$	16 100	~12	~ 4
$\{Pd_2(dmb)_2(dpa)^{2+}\}_n$	19 500	~16	~ 5
$\{Pd_2(dppe)_2(dmb)^{2+}\}_n$	11 800	~ 8	~ 2
$\{Pd_2(dppp)_2(dmb)^{2+}\}_n$	12 100	~ 8	~ 2
$\{Pd_2(dppp\mathbf{R})_2(dmb)^{2+}\}_n$	13 600	~ 7	~ 2

Even more interesting, the number of units evaluated from the T_1/NOE method is systematically 3 – 4 times smaller than that measured from the intrinsic viscosity properties. This discrepancy is due to the presence of chain flexibility, which allows "local" molecular motions. The probe ^{13}C nuclei senses motions that are the equivalent to 4-5 units for the "red" series, and about only 2 for the "yellow" series (dmb being sufficiently flexible for that). In support of this explanation, this method was recently applied to unambiguously confirmed mass data that suggested dimeric structures for a series of rhodium(I) complexes of diphosphinated-calix[4]arenes.[14] Molecular modeling showed very rigid dimers due to the presence of slightly encumbered rings. In these cases, and the $\{Ag(dmb)_2^+\}_n$ cases presented above, the rigidity of the molecular structures allowed the accurate estimation of the molecular dimension by T_1/NOE.

Coming back to molecular dynamics, by changing the flexible alkyl chain for a more rigid group (from $-(CH_2)_m-$ to $-C\equiv C-$), or by adding large lateral chains to the backbone ($O-(CH_2)_6-O-$Naphthyl), no sensitive effect is observed on the estimated number of units by T_1/NOE. These results indicate that the presence of lateral chains do not affect the molecular motions about the ^{13}C probe, and that the $-C\equiv C-$ group is still fairly flexible due to the absence of an important barrier to rotation.

Conclusion

The T_1/NOE method is not new. Ironically, it was used long ago for the estimation of the molecular weight of proteins. The dimensions of proteins are generally so large that the slow narrowing limit must apply ($\omega\tau_c \gg 1$), and the more complex Equation 2 must be used. In addition, accuracy is best achieved when a comparison with a known standard is made. Finally, chain flexibility represents an important problem in the application of this method. Considering these facts, some earlier studies may need to be revisited. For this research program, the application of this method works best for oligomers, which exhibit M_n that are often a problem for most techniques in polymer science. The low solubility of some materials can also be overcome by simply accumulating data for a longer period of time.

Acknowledgments

This research was supported by the NSERC (Natural Sciences and Engineering Research Council of Canada). PDH thanks the graduate students who did all the work; D. Perreault, D. Fortin, N. Jourdan, T. Zhang, M. Turcotte, F. Lebrun, E. Fournier, S. Sicard, P. Mongrain, and J.-F. Fortin.

[1] F. W. Wehrli, T. Wirthlin, *"Interpretation of Carbon-13 NMR Spectra"*, Heyden, London 1980.

[2] R. J. Abaham, P. Loftus, *"Proton and Carbon-13 NMR Spectroscopy; An Integrated Approach"*, Heyden, London 1980.

[3] M. Turcotte, P. D. Harvey, *Inorg. Chem.* **2002**, *41*, 1739.

[4] M. Turcotte, M.Sc. Dissertation, Université de Sherbrooke, 2000.

[5] J.-M. Basset, D. E. Berry, G. E. Barker, M. Green, J. A. K. Howard, F. G. A. Stone, *J. Chem. Soc., Dalton Trans.* **1979**, 81.

[6] D. Perreault, M. Drouin, A. Michel, P. D. Harvey, *Inorg. Chem.* **1992**, *31*, 3688.

[7] D. Fortin, M. Drouin, M. Turcotte, P. D. Harvey, *J. Am. Chem. Soc.* **1997**, *119*, 531.

[8] T. Zhang, M. Drouin, P. D. Harvey, *Inorg. Chem.* **1999**, *38*, 957.

[9] T. Zhang, M. Drouin, P. D. Harvey, *Inorg. Chem.* **1999**, *38*, 1305.

[10] F. Lebrun, M.Sc. Dissertation, Université de Sherbrooke, **2001**.

[11] D. Fortin, M. Drouin, P. D. Harvey, *J. Am. Chem. Soc.,* **1998**, *120*, 5351.

[12] D. Fortin, M. Drouin and P.D. Harvey, *Inorg. Chem.* **2000**, *39*, 2758.

[13] S. Sicard, M.Sc. Dissertation, Université de Sherbrooke, **2003**.

[14] F. Plourde, K. Gilbert, J. Gagnon, P. D. Harvey, *Organometallics* **2003**, *22*, 2866.

Macromol. Symp. **2004**, *209*, 81-95

Luminescence Properties of Organometallic/Coordination Oligomers and Polymers Containing Diphosphine and Diisocyanide Assembling Ligands: Comparison between Mononuclear Model Complexes and Polymers

Pierre D. Harvey

Département de chimie, Université de Sherbrooke, Sherbrooke, PQ, Canada, J1K 2R1
E-mail: pierre.harvey@usherbrooke.ca

Summary: This paper describes the luminescence properties of selected examples of homo- and mixed-bridging ligand-containing organometallic/coordination oligomers and polymers of copper(I), silver(I), palladium(I) and –(0.5), and platinum(0.5). The bridging ligands are the 1,8-diisocyano-p-menthane (dmb) and bis(diphenylphosphino)methane (dppm), -ethane (dppe), -propane (dppp), -butane (dppb), -pentane (dpppen), and –hexane (dpph), as well as bis(diphenyl-phosphino)acetylene (dpa) and bis(dimethylphosphino)methane (dmpm). The comparison between the mononuclear model complexes and the polymers of the emission maxima (λ_{max}) and emission lifetimes (τ_e) will be made, and interpreted by the presence of intrachain interactions and exciton phenomena.

Keywords: copper; diisocyanides; diphosphines; emission lifetimes; exciton; luminescence; palladium; platinum; photophysics; silver

Introduction

The areas of organometallic/coordination polymers as well as crystal engineering and supramolecular assemblies have become extremely active in the past decade. Applications in the fields of supported catalysis, non-linear optics (NLO), light emitting diodes (LED), conducting and photoconducting materials are among the few examples of motivation for these researches. The luminescence properties are important for

 DOI: 10.1002/masy.200450506

sensorizations and display devices, as well as LED applications, and their report in the literature are frequent (see recent examples on copper (I), silver (I), and platinum(0) see ref.[1-17]). However, their reports focus primarily on the luminescence spectroscopic and photophysical characterizations of the materials. In depth investigations on what effect their luminescence properties have, and on the excited state relaxation mechanism, are rather scarce.

We now wish to report a series of results that address the difference between the photophysical properties of the mononuclear complexes used as models for a single unit of a polymer, versus the corresponding polymers or oligomers. The list of bridging ligands is shown in Scheme 1.

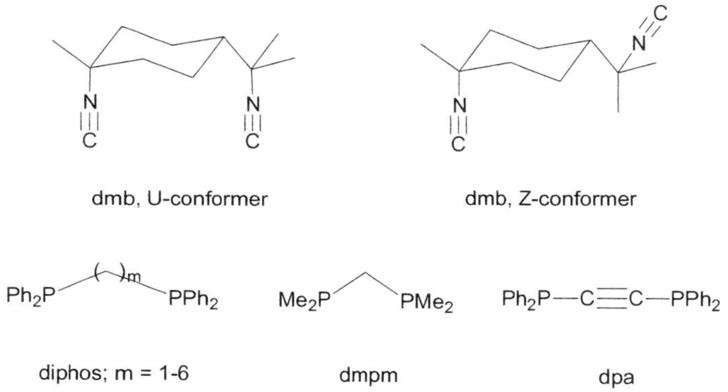

dmb, U-conformer   dmb, Z-conformer

diphos; m = 1-6   dmpm   dpa

Scheme 1

Background

In 1992, this group reported a new type of coordination/organometallic polymer based on silver(I) and bridging ligand dmb.[18] The cationic polymers exhibit a 1-D structure and is of the type $\{Ag(dmb)_2]Y\}_n$ (Y$^-$ = BF$_4^-$, PF$_6^-$, NO$_3^-$) where the dmb ligand adopts the U-conformer. The result was unprecedented because there was, no precedent for the preparation of polymers using this ligand, as recently reviewed.[19] Polymers of the type $\{Ag(Z-dmb)_2]Y\}_n$ and $\{Ag(U-dmb)(Z-dmb)]Y\}_n$ where Y$^-$ = TCNQ$^-$ (teracyano-quinodimethane anion) have also been reported.[20] The silver atoms in $\{Ag(U-dmb)_2]Y\}_n$

$(Y^- = BF_4^-, PF_6^-, NO_3^-)$ are tetracoordinated by four isocyanide groups forming a distorted geometry where the CAgC angles deviate strongly from the ideal tetrahedral angle of 109.45° according to X-ray for the PF_6^- salt, presumably due to intraligand steric hindrance.[19] The Ag···Ag distance is about 5 Å, and the Ag_3 angle is 140°. Subsequently, the Cu analogues were prepared and fully characterized, and the photophysical properties were also reported in detail as well.[21] Both types of polymers, Cu- and Ag-, are strongly luminescent in the solid states and in solution at 77 K in ethanol for example (λ_{max}(Cu) = 548 nm; λ_{max}(Ag) = 502 nm). The key feature is that the emission bands are very broad (FWHM = 150 nm) and the plots of ln of the luminescene intensity after an excitation pulse as a function of delay time (i.e. emission decay curves) are not linear, which are normally typical for unimolecular processes. The fact that the decay curves superimposed very well for solid state and solution data indicated that this unusual property was due to an intrachain phenomena. This behavior was identified as an exciton property,[22] and triggered an exhaustive investigation of the photophysical properties of organometallic /coordination polymers of copper(I), silver(I), palladium(I) and –(0.5), and platinum(0.5) (Scheme 2).

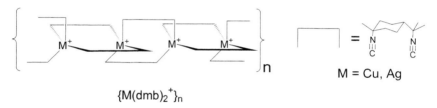

$\{M(dmb)_2^+\}_n$

Scheme 2

Emission Data Analysis

The emission lifetimes can be analyzed in two ways. The first way was the common deconvolution method allowing for analysis of the raw data using models from 1 to 4 exponential decays. For unimolecular processes, the plots are linear and the slopes are related to the lifetimes (slope = $-1/\tau_e$). For multi-processes (bi-, tri- and tetra-exponentials), the results provide both the lifetimes and the relative weight for each component. The second way is the ESM (Exponential Series Method),[23, 24] and consists of calculating a decay curve that may be composed of many components (up to 200 instead of only 4) with a window of calculated lifetimes well exceeding the time window of the data, insuring that any short or long components can be detected. Again the results are presented as lifetimes and relative weights, and the plots of lifetime versus weight represent the lifetime distribution. A single lifetime event gives a very narrow distribution. The quality of the fit between the experimental and calculated curves are also addressed using parameter χ (goodness-of-fit), and from the residual analysis. A good fit is accompanied with a residual where the "noise" above and under the baseline is evenly distributed.

The Polymers of Bi- and Tetranuclear Clusters

In 1997, Zhang et al reported the syntheses and characterization of new tetranuclear clusters of palladium and platinum formulated as $M_4(dmb)_4(PPh_3)_2^{2+}$ (M = Pd, Pt).[25, 26] Both types were unique because for palladium it was the first time that a 58-valence electron palladium cluster was reported. In addition, this M_4-cluster was linear, and a catenate sub-structure involving two interlocking "$Pd_2(dmb)_2$" rings using three Pd-Pd bonds (1 inner, and 2 outer), occurred (Scheme 3). Similarly for platinum, the same unprecedented features were noted, except that some other rare example of 58-electron species were already known. The X-ray structures for both species were reported, and the M-M bonds are as follows: for $Pd_4(dmb)_4(PPh_3)_2^{2+}$, d(Pd-Pd) = 2.534(13) inner, and 2.524(10) Å outer; for $Pt_4(dmb)_4(PPh_3)_2^{2+}$, d(Pt-Pt) = 2.641(2) inner, and 2.654(2) and 2.666(2) Å outer. For both cases, polymers of clusters have been prepared and fully characterized using crystallographic (for the red zig-zag shape $\{Pd_4(dmb)_4(Z-dmb)^{2+}\}_n$) and spectroscopic methods, and using molecular weight measurements (intrinsic viscosity for the amorphous orange $\{Pt_4(dmb)_4(diphos)^{2+}\}_n$; diphos = dppb, MW_n = 203 000; dpppen, MW_n = 307 000; dpph, MW_n = 84 000).

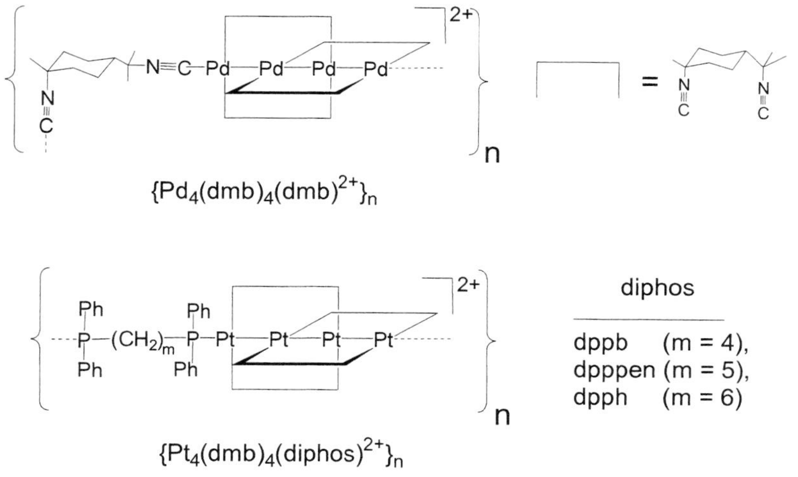
2+
N≡C–Pd–Pd–Pd–Pd
N
C
n
=
{Pd4(dmb)4(dmb)2+}n
Ph
P–(CH2)m–P–Pt–Pt–Pt–Pt
Ph
2+
n
diphos
dppb (m = 4),
dpppen (m = 5),
dpph (m = 6)
{Pt4(dmb)4(diphos)2+}n

Scheme 3

Both series of polymers exhibit a lowest energy excited state of the type dσ→dσ* and are luminescent at 77K (Table1). Numerous comparisons can be made. First, the λ_{max} appears to be dependent upon whether it is a polymer versus the Pd_4-cluster species. The Pd_4-polymer structureless emission band is slightly more red-shifted by about 15 nm than the Pd_4-cluster. This result may reflect the difference in electron donating ability of the isocyanide versus phosphine, where the alkyl isocyanide is a stronger electron donor that PPh_3. On the other hand, such a comparison is not possible in the Pt_4-containing series. Variations are observed, but the phosphines are relatively similar. However, a red-shift of the emission band is observed going from dppb to dpppen to dpph. This trend may reflect that the CH_2-containing groups, more electron donating the phosphine, is. Some solvent effect is noted, where the use of PrCN leads to a red-shift of the emission maximum, as well as a decrease in lifetimes. Perhaps the most interesting feature is the change in emission lifetimes going from the cluster to its corresponding polymers; an approximate 2-fold increase is noted for both series. This increase is unexpected when considering the well-known "loose bolt" effect, where an increase in excited state deactivation is anticipated due to an increase in internal conversion associated with molecular vibration.[22] This result is better explained by the restriction in molecular motion of the chromophore. When the cluster is free to move, then excited state deactivation is enhanced. On the other hand, when the cluster is attached at both ends, the free rotation and torsion are much more restricted, leading to a lesser extent to excited deactivation.

Table 1. Photophysical data for the $M_4(dmb)_4{}^{2+}$-containing species.

Materials[a]	λ_{max}	τ_e	Conditions
	nm	ns	
$[Pd_4(dmb)_4(PPh_3)_2]^{2+}$	684	0.67 ± 2	77K/EtOH
$[Pd_4(dmb)_4(PPh_3)_2]^{2+}$	685	0.56 ± 1	77K/PrCN
$\{[Pd_4(dmb)_4(Z\text{-}dmb)]^{2+}\}_n$	698	1.14 ± 5	77K/EtOH
$\{[Pd_4(dmb)_4(Z\text{-}dmb)]^{2+}\}_n$	703	1.00 ± 2	77K/PrCN
$[Pt_4(dmb)_4(PPh_3)_2]^{2+}$	750	2.71 ± 5	77K/EtOH
$\{Pt_4(dmb)_4(dppb)\}_n$	736	4.78 ± 5	77K/EtOH
$\{Pt_4(dmb)_4(dpppen)\}_n$	750	5.15 ± 5	77K/EtOH
$\{Pt_4(dmb)_4(dpph)\}_n$	755	5.17 ± 5	77K/EtOH

Recently, a second and original series of new polymers of polynuclear palladium series has been reported.[27-29] This series is divided into two series: 1) $\{Pd_2(dmb)_2(diphos)^{2+}\}_n$ (diphos = dppb, dpppen, dpph, and dpa), which are red,[27, 28] and 2) $\{Pd_2(diphos`)_2(dmb)^{2+}\}_n$ (diphos` = dppe, dppp), which are yellow. Both series are prepared from the d^9-d^9 binuclear complex $Pd_2(dmb)_2Cl_2$ (Scheme 4).[28, 29] The materials in the solid state are amorphous based upon XRD measurements, and no X-ray crystallographic data could be obtained, despite numerous attempts. Based upon the intrinsic viscosity measurements, the MWn for $\{Pd_2(dmb)_2(diphos)^{2+}\}_n$ are 17 800 (dppb, ~14 units), 18 400 (dpppen, ~14 units), 16 100 (dpph; ~12 units), and 19 500 (dpa, ~ 12 units), and for $\{Pd_2(diphos`)_2(dmb)^{2+}\}_n$ the MWn are 11800 (dppe, ~8 units) and 12 100 (dppp, ~8 units).

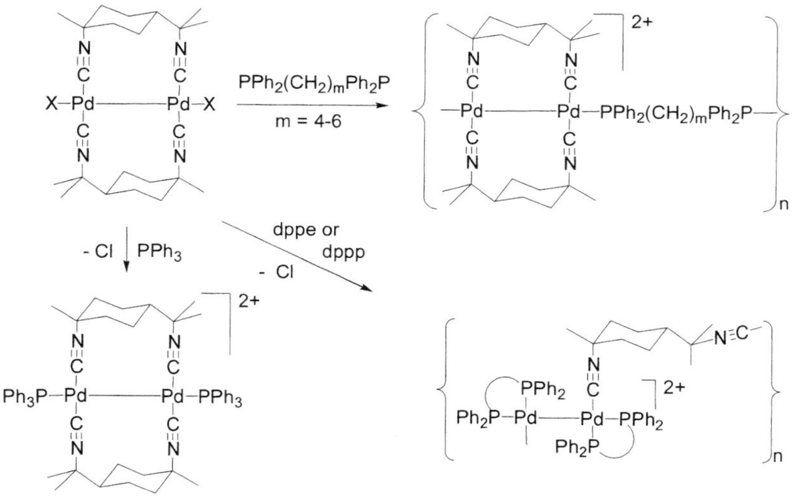

Scheme 4

The complex $Pd_2(dmb)_2(PPh_3)_2^{2+}$ can be used as a comparison molecule. The lowest energy excited state is also of the type $d\sigma \rightarrow d\sigma^*$. This complex and the oligomers are not luminescent in solution nor in the solid state at room temperature, and may be associated with an energy wasting photo-induced hemolytic Pd_2-bond scission or ligand dissociation in the excited state. However, the compounds become luminescent at 77 K (in PrCN). The $Pd_2(dmb)_2(PPh_3)_2^{2+}$ complex and the $\{Pd_2(dmb)_2(diphos)^{2+}\}_n$ oligomers (diphos = dppb, dpppent, dpph, dpa) exhibit emission maxima (λ_{max}) between 627 and 638 nm (Table 2). The excitation spectra superpose the absorptions relatively well, indicating that the absorbing and emitting species are the same. The emission λ_{max} data are more red-shifted in comparison with those observed for the $\{Pd_2(diphos`)_2(dmb)^{2+}\}_n$ oligomers ($500 < \lambda_{max} < 509$ nm), which is consistent with the more blue-shifted $d\sigma \rightarrow d\sigma^*$ absorptions of the latters (diphos`= dppe (414), dppp (400)). The Stokes shifts (difference between the absorption and emission maxima; Δ) are in the order of 4 500 to 5 600 cm^{-1}, suggesting that the luminescence is phosphorescence. However, the ns timescale is strikingly short, but not unprecedented. For example, the luminescence of the related

$Pd_2(dmb)_2Cl_2$ dimer exhibits a τ_e of 71 ± 6 ns at 77 K.[30] On the other hand, the ns timescale is also too long to be assigned to a fluorescence (the fluoresecence lifetime for 4d- and 5d-containing chromophore is generally in the ps time scale). The former assignment is preferred. The emission quantum yields appear constant for most species ($0.12 < \Phi < 0.17$), which is consistent with the similarity in τ_e data, but three of them listed in Table 2 (diphos = dppb, dpppen, dpph) exhibit 2 to 8 fold decreases. This difference is unexplained at the moment.

Table 2. Spectroscopic and photophysical data for the oligomers in PrCN at 77K.

Compounds	λ_{max}	Δ	τ_e	Φ
	nm	± 100 cm^{-1a}	± 0.08 ns	$\pm 10\%$
$[Pd_2(dmb)_2(PPh_3)_2](ClO_4)_2$	627	4 900	2.75	0.14
$\{[Pd_2(dmb)_2(dppb)](ClO_4)_2\}_n$	632	5 100	1.87	0.026
$\{[Pd_2(dmb)_2(dppen)](ClO_4)_2\}_n$	634	5 200	2.70	0.071
$\{[Pd_2(dmb)_2(dpph)](ClO_4)_2\}_n$	636	4 900	2.24	0.046
$\{[Pd_2(dmb)_2(dpa)](ClO_4)_2\}_n$	638	4 800	2.30	0.15
$\{[Pd_2(dppet)_2(dmb)](ClO_4)_2\}_n$	509	4 500	1.94	0.13
$\{[Pd_2(dpppro)_2(dmb)](ClO_4)_2\}_n$	508	5 300	1.50	0.12

The 1-D dmb-Polymers of Copper(I) and Silver(I)[21]

A brief introduction on the $\{M(dmb)_2^{2+}\}_n$ polymers (M = Cu; Ag) is presented above,[18-21] and further relevant information on their properties can be found in the following ref.[31-34] The emission spectra are broad (~150 nm) and centered around 550 and 500 nm for M = Cu and Ag, respectively. The graphs of the ln of luminescence intensity as a function of delay time after the light pulse (decay curve) is not mono-exponential. It is, in fact, non-exponential. An examination of the emission spectra as a function of delay time (time-resolved emission spectroscopy) indicates the presence of an "infinite" number of discrete emission bands contributing to the overall spectral envelope. By comparing the slopes in the decay curves and the λ_{max} of emission in the time-resolved spectra at the early event after the pulse with those obtained for the mononuclear complexes $M(CN\text{-}t\text{-}Bu)_4^+$ (M = Cu, Ag) as model compounds for a single chromophore unit, one readily depicts the close resemblance. This important information indicates that at the early event after the light pulse, the chromophore $M(CN\text{-}t\text{-}Bu)_4^+$ absorbs the light and luminesces as a "non-interacting species". As time evolves, other and slower photophysical events occur with red-shifted emissions. Because of the fact that the decay traces are almost identical for the polymers in the solid state and in solution, one come to the conclusion that this photophysical process occurs within the chain. Because of this multiple emission behavior, it is not surprising that the emission light is found depolarized. This process is known as an exciton phenomenon (Scheme 5).

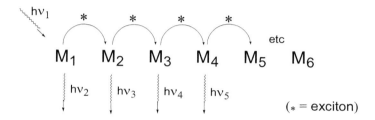

Scheme 5

Based on DFT calculations, the lowest energy excited state of the $M(CN\text{-}t\text{-}Bu)_4^+$ chromophore is a metal-to-ligand-charge-transfer (MLCT). Based on the X-ray data of the $\{Ag(dmb)_2^{2+}\}_n$ polymers, the N⋯N distance is in the range of 4.3 Å,[33] a distance that is close to the sum of the van der Waals radii (3.4-3.6 Å), and sufficiently short for weak interactions and energy transfer to occur. Normally, exciton phenomena are important in the singlet state for organics because of the selection rule, but in the triplet state, this property should not be important. In fact, for metal-containing species, spin-orbit coupling renders the selection rule less rigorous, and such a process becomes possible. One of the key features is the knowledge of the molecular dimension of the $\{M(dmb)_2^{2+}\}_n$ polymers in solution (M = Cu, number of units = 300;[21] M= Ag, number of units = 8).[34] The fact that the X-ray structure for $\{Ag(dmb)_2^{2+}\}_n$ demonstrates the

90

presence of a polymer, not an oligomer of 8 units, is also striking because the photophysical data are almost identical, as stated above. This observation indicates that energy transfer is reversible (Scheme 6), which is consistent with the fact that the chromophores are identical (no donor and acceptor). The questions are now to what extent this energy delocalisation takes places, and how can they be fine-tuned.

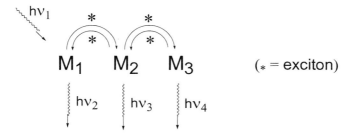

Scheme 6

The Small dmb-Oligomers of Copper(I) and Silver(I)

In an attempt to control the molecular dimension of the interacting Cu- and Ag-containing chromophores, small oligomers have been synthesized.[35] Indeed, the luminescent binuclear complexes $M_2(dmpm)_3^{2+}$ (M = Cu, Ag), $Ag_2(dmpm)_2^{2+}$ and $Cu_2(dmpm)_3(CN\text{-}t\text{-}Bu)_2^{2+}$ (as BF_4^- salts), as well as the oligomers described as $\{Cu_2(dmpm)_3(dmb)_{1.33}^{2+}\}_3$ and $\{Ag_2(dmpm)_2(dmb)_{1.33}^{2+}\}_3$, have been prepared and fully characterized in the solid state. These binuclear compounds and small oligomers bear the chromophore units $M_2(dmpm)_3^{2+}$, $Ag_2(dmpm)_2^{2+}$, $M_2(dmb)_2^{2+}$ (M = Cu, Ag) and $Ag_2(dmb)^{2+}$ (Scheme 7), and exhibit emission maxima ranging from 445 to 485 nm with emission lifetimes found in the µs regime.

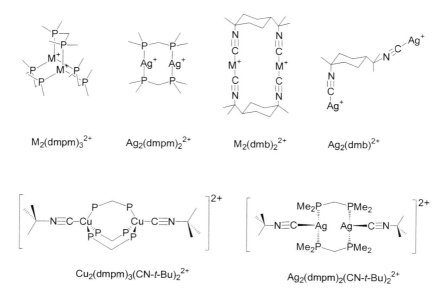

$M_2(dmpm)_3^{2+}$ $Ag_2(dmpm)_2^{2+}$ $M_2(dmb)_2^{2+}$ $Ag_2(dmb)^{2+}$

$Cu_2(dmpm)_3(CN\text{-}t\text{-}Bu)_2^{2+}$ $Ag_2(dmpm)_2(CN\text{-}t\text{-}Bu)_2^{2+}$

Scheme 7

The time-resolved emission spectra for the oligomers and the known polymers $\{M(dmb)_2^+\}_n$ (M = Cu, Ag) also exhibit blue-shifted emission bands at the early stage of the photophysical event after the light pulse, which also red-shift with longer delay times (Figure 1).

The decay traces are more or less exponential as well exemplified shown in Figure 2. In this graph, $[Cu_2(dmpm)_3(CN\text{-}t\text{-}Bu)_2](BF_4)_2$ exhibits a relatively linear decay, indicating the poor extent of exciton delocalisation. On the other hand, the decay trace for the $\{Cu(dmb)_2^{2+}\}_n$ polymer (here 45 units), exhibits a non-exponential behavior.

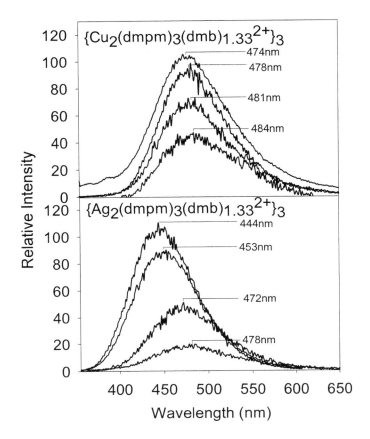

Figure 1. Time-resolved emission spectra for solid $\{Cu_2(dmpm)_3(dmb)_{1.33}^{2+}\}_3$ (above) and $\{Ag_2(dmpm)_2(dmb)_{1.33}^{2+}\}_3$ (below) at room temperature. The measurements have been made in the following time frames: for above: 474 nm, 20-70; 478, 500-600; 481, 1 000-1 300; 484; 2 000-2 500μs; for below: 444 nm, 20-70; 453, 300-400; 472, 500-600; 478, 1 000-1 300 μs.

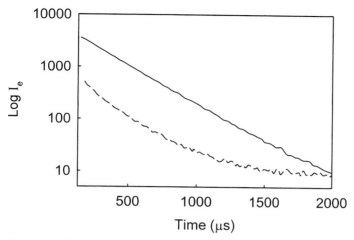

Figure 2. Solid state decay traces for the emission of $[Cu_2(dmpm)_3(CN\text{-}t\text{-}Bu)_2](BF_4)_2$ (——) versus $\{Cu(dmb)_2^+\}_n$ (----) at 298K.

These non-exponential decay traces were analyzed by ESM and exhibit a distribution of lifetimes that can be fairly broad as exemplified in Figure 3. It is interesting to note that the width of the distribution is a function of the number of units in the chain. The comparison is crude because the nature of the emissive excited state is not the same for all three examples. For $Ag_2(dmpm)_2^{2+}$, the lowest excited state is $d\sigma^*\text{-}p\sigma$,[36-42] not MLCT. So the electronic transition is confined within the M_2P_6 or M_2P_4 unit, which are of a smaller dimension in comparison with "$M(CN)_x$" chromophores. Hence, the shortest interchromophore distance may not be the same from one system to another. The extent of red-shift of the emission band with delay time and the fwhm (full-width-at-half-maximum) of the distribution of lifetimes also appears to increase as the number of M_2-units increases in the oligomers. This behavior is also consistent with an exciton process, as described above.

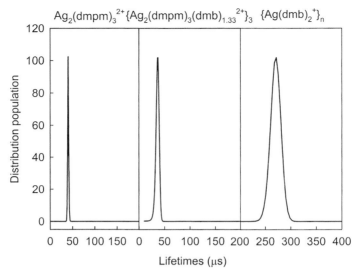

Figure 3. Comparison of the distribution of lifetimes fitting emission decay traces for solid $Ag_2(dmpm)_2^{2+}$, $\{Ag_2(dmpm)_2(dmb)_{1.33}^{2+}\}_3$ and $\{Ag(dmb)_2^+\}_n$ (n = 8) at 298K.

Conclusion

This paper focused on demonstrating that there is a clear difference between the properties of discrete complexes and organometallic/coordination polymeric materials. Evidence for inter-unit interactions, steric or excitonic, is present, and these properties clearly modulate the overall optical properties of the bulk. Knowing this, fine tuning of the photophysical properties for NLO, LED and digital display applications appears possible in this way.

Acknowledgments

This research was supported by the NSERC (Natural Sciences and Engineering Research Council of Canada). PDH thanks the graduate students who did all the work; D. Perreault, D. Fortin, N. Jourdan, T. Zhang, M. Turcotte, F. Lebrun, E. Fournier, S. Sicard, and J.-F. Fortin.

[1] X.-M. Ouyang, Z.-W. Li, T.-A. Okamura, Y.-Z. Li, W.-Y. Sun, W.-X. Tang, N. Ueyama, *J. Solid State Chem.* **2004**, *177*, 350.
[2] C. Seward, J. Chan, D. Song, S. Wang, *Inorg. Chem.* **2003**, *42*, 1112.
[3] Y.-B. Dong, G.-X. Jin, M. D. Smith, R.-Q. Huang, B. Tang Bo, H.-C. Zur Loye, *Inorg. Chem.* **2002**, *41*, 4909.
[4] M.-L. Tong, J.-X. Shi, X.-M. Chen, *New J. Chem.* **2002**, *26*, 814.
[5] S.-L. Zheng, M.-L. Tong, S.-D. Tan, Y. Wang, J.X. Shi, Y.-X. Tong, H. K. Lee, X.-M. Chen, *Organometallics* **2001**, *20*, 5319.
[6] X.-H. Bu, H. Liu, M. Du, K. M.-C. Wong, V. W.-W. Yam, M. Shionoya, *Inorg. Chem.* **2001**, *40*, 4143.
[7] M.-L. Tong, X.-M. Chen, B.-H. Ye, L.-N. Ji, *Angew. Chem., Int. Ed.* **1999**, *38*, 2237.
[8] T. Yasuda, T. Yamamoto, *Macromolecules* **2003**, *36*, 7513.
[9] C. Nather, J. Greve, I. Jess, C. Wickleder, *Solid State Sci.* **2003**, *5*, 1167.
[10] R.-Z. Li, D. Li, X.-C. Huang, Z.-Y. Qi, X.-M. Chen, *Inorg. Chem. Commun.* **2003**, *6*, 1017.
[11] Y.-B. Dong, J.-Y. Cheng, H.-Y. Wang, R.Q. Huang, B. Tang, M. D. Smith, H.-C. Zur Loye, *Chem. Materials* **2003**, *15*, 2593.
[12] J. Zhang, R.-G. Xiong, X.-T. Chen, Z. Xue, S.-M. Peng, X.-Z. You, *Organometallics* **2002**, *21*, 235.
[13] J. Zhang, R.-G. Xiong, X.-T. Chen, C.-M. Che, Z. Xue, X.-Z. You, *Organometallics* **2001**, *20*, 4118.
[14] N. W. Alcock, P. R. Barker, J. Haider, M. J. Hannon, C. L. Painting, Z. Pikramenou, E. A. Plummer, K. Rissanen, P. Saarenketo, *J. Chem. Soc., Dalton Trans.* **2000**, 1447.
[15] S. Parsons, Z. Pikramenou, G. A. Solan, R. E. P. Winpenny, *J. Cluster Sci.*, **2000**, *11*, 227.
[16] A. Kokil, C. Huber, W. R. Caseri, C. Weder, *Macromol. Chem. Phys.* **2003**, *204*, 40.
[17] C. Huber, F. Bangerter, W. R. Caseri, C. Weder, *J. Am. Chem. Soc.* **2001**, *123*, 3857.
[18] D. Perreault, M. Drouin, A. Michel, P. D. Harvey, *Inorg. Chem.* **1992**, *31*, 3688.
[19] P. D. Harvey, *Coord. Chem. Rev.* **2001**, *219-221*, 17.
[20] D. Fortin, M. Drouin, P.D. Harvey, *J. Am. Chem. Soc.* **1998**, *120*, 5351.
[21] D. Fortin, M. Drouin, M. Turcotte, P. D. Harvey, *J. Am. Chem. Soc.* **1997**, *119*, 531.
[22] N. J. Turro, "*Modern Molecular Photochemistry*", Benjamin Cummings, Menlo Park 1978, p 354.
[23] A. Siemiarczuk, B. D. Wagner, W. R. Ware, *J. Phys. Chem.* **1990**, *94*, 1661.
[24] A. Siemiarczuk, W. R. Ware, *Chem. Phys. Lett.* **1989**, *160*, 285.
[25] T. Zhang, M. Drouin, P.D. Harvey *Inorg. Chem.* **1999**, *38*, 957.
[26] T. Zhang, M. Drouin, P.D. Harvey *Inorg. Chem.* **1999**, *38*, 1305.
[27] S. Sicard, F. Lebrun, J.-F. Fortin, A. Decken, P.D. Harvey, *Inorg.Chem.* **2004**, submitted.
[28] S, Sicard, M.Sc. Dissertation, Université de Shebrooke, 2004.
[29] É. Fournier, S. Sicard, F. Lebrun, A. Decken, P.D. Harvey, *Inorg. Chem.* **2004**, in press.
[30] P. D. Harvey, Z. Murtaza, *Inorg. Chem.*, **1993**, *32*, 4721.
[31] D. Fortin, P.D. Harvey, *Coord. Chem. Rev.* **1998**, *171*, 351.
[32] D. Fortin, M. Drouin, P.D. Harvey, F.G. Herring, D.A. Summers, R.C. Thompson, *Inorg. Chem.* **1999**, *38*, 1253.
[33] D. Fortin, M. Drouin, P.D. Harvey, *Inorg. Chem.* **2000**, *39*, 2758.
[34] M. Turcotte, P. D. Harvey, *Inorg. Chem.* **2002**, *41*, 1739.
[35] É. Fournier, M.Sc. Dissertation, Université de Shebrooke, 2003.
[36] K. H. Leung, D. L. Phillips, Z. Mao, C.-M. Che, V. M. Miskowski, C.-M. Chan, *Inorg. Chem.* **2002**, *41*, 2054.
[37] H.-X. Zhang, C.-M. Che, *Chem. Eur. J.* **2001**, *7*, 4887.
[38] W.-F. Fu, K.-C. Chan, K.-K. Cheung, C.-M. Che, *Chem. Eur. J.* **2001**, *7*, 4656.
[39] K. H. Leung, D. L. Phillips, M.-C. Tse, C.-M. Che, V. M. Miskowski, *J. Am. Chem. Soc.* **1999**, *121*, 4799.
[40] W.-F. Fu, K.-C. Chan, V. M. Miskowski, C.-M. Che, *Angew. Chem. Int. Ed.* **1999**, *38*, 2783.
[41] P. D. Harvey, H. B. Gray, *J. Am. Chem. Soc.* **1988**, *110*, 2145.
[42] P. D. Harvey, R. F. Dallinger, W. H. Woodruff, H. B. Gray, *Inorg. Chem.* **1989**, *28*, 3057.
[43] D. Piché, P. D. Harvey, *Can. J. Chem.* **1994**, *72*, 705.

Metallodendritic Materials for Heterogenized Homogeneous Catalysis

Olivier Bourrier, Ashok K. Kakkar*

Department of Chemistry, McGill University, 801 Sherbrooke St. West, Montreal, Quebec, H3A 2K6 Canada
E-mail: ashok.kakkar@mcgill.ca

Summary: Design of supports containing a hyperbranched backbone and active transition metal centers at the periphery is described. Such nanoarchitectures can be easily assembled from 3,5-dihydroxybenzyl alcohol and dimethylsilylamine using a divergent synthetic methodology. Following a controlled reaction pathway, construction of dendrimers of up to generation 5 is achieved, while a simple mix of reagents in one-pot or with sequential additions yields analogous hyperbranched polymers. Subsequent functionalization at the periphery with phosphine followed by Rh(I) centers yields the desired metallodendritic materials. The efficiency of the latter in catalytic hydrogenation of decene under varied conditions is surveyed.

Keywords: dendrimers; hyperbranched polymers; organometallic dendrimers; supported metal catalysis

Introduction

Ideally, a catalytic system should be well defined, enable rapid and selective chemical transformations, and its complete separation from the reaction mixture should easily be achieved.[1-4] Catalysts fall into two classes based on their physical behaviour in solution. They are referred to as "homogeneous" when they are solvated with the reactants. Homogeneous catalysis offers an easy access to all catalytic moieties, and reactions can be done in a high metal-to-substrate ratio.[5] However, their separation from the products is difficult and remains a major issue. Common methods of separation include distillation, liquid-liquid extraction, crystallization or even destruction of the catalyst.[5] "Heterogeneous" catalysts are insoluble materials by definition. They offer the advantages such as high thermal and mechanical stability, a loading potential far superior than the homogeneous catalysts, and of course an inherent ability to be easily separated from the product.[6] However, their nonuniformity and ill-defined structures tend to limit

© 2004 WILEY-VCH Verlag GmbH & KGaA, Weinheim   DOI: 10.1002/masy.200450507

98

the accessibility to the catalytic sites. Other problems encountered with heterogeneous catalysts are related to slow diffusion and metal leaching, hence a lower catalytic activity when compared to their homogeneous analogs.

Dendrimers appear to offer a good compromise between the homogeneous and heterogeneous catalysts. Dendrimers are well defined materials with structure and size that can be tuned to support or host transition metal catalysts, and the location as well as number of catalytic sites can be regulated.[7,8] The globular shapes they adopt in higher generations seem particularly suited to new means of catalyst-product separation such as the nanofiltration using a membrane.[1] Seminal work on the first examples of metallodendrimers used in catalysis is attributed to van Koten's group with the preparation of a dodeca(aryl bromide) terminated carbosilane dendrimers.[4] In this study, peripheries of carbosilane dendrimers were functionalized by oxidative addition of [Ni(PPh₃)₄] to afford diaminoarylnickel(II) terminated dendrimer. The resulting metallodendrimers were used to catalyze the Kharasch addition of tetrachloromethane to C=C bonds. Turnover frequencies were found to be 30% lower than the monomeric or polymer-bound analogs.

The design and topology of metallodendrimers can define the activity and selectivity of the resulting catalyst.[7-10] The catalytically active metal centers could be present at the periphery of hyperbranched dendrimers, or of star branch dendrimers. Transition metals can also be part of the backbone, or found at the center, acting as a core. They can be coordinated throughout the backbone, as host. Also, periphery metal supported dendritic wedge can be grafted on a polymeric support.

The majority of supported metal dendrimer catalysts are the result of the peripheral modification of pre-existing dendritic macromolecules. Although an attractive route, this can lead to a decrease in catalytic activity by steric congestion if the concentration of metal centers becomes too high. For example, van Koten and co-workers[11] reported the preparation of periphery palladated dendrimer using the same carbosilane backbone as that of his Ni metallodendrimers.[4] Catalytic activity of the supported Pd dendrimers decreased as the steric congestion increased, as observed for the Ni metallodendrimers. Other examples of catalytically active metallodendrimers can be found in the literature.

Peripheral functionalization with varied transition metal centers has been achieved in star branched dendrimers[12] as well as hyperbranched dendrimers.[13-18]

Hoveyda et al.[12] reported the preparation of recyclable poly(organoruthenium) metathesis star carbosilane catalysts. Ring closing metathesis of bis(allyl)-*N*-tosylamide to the corresponding cyclopentane was reported using the latter dendrimers with yields exceeding 90%, and only a slight loss of Ru upon recovery.

P-branched dendrimers synthesized by Majoral et al.[13] have been functionalized with a variety of transition metals coordinated to terminal phosphine ligands.[14,15] $PdCl_2$ terminated dendrimers were used to catalyze the Stille coupling reaction whereas the $RuH_2(PPh_3)_2$ terminated dendrimers catalyzed the Knoevenagel condensation and the Michael addition reactions.[16]

Also, phosphanyl-terminated dendrimers have been prepared by Reetz et al.[17] from third generation poly(propylene imine) dendrimer treated with Ph_2PCH_2OH. Treatment with $[PdCl_2(PhCN_2)]$, $[Pd(CH_3)_2(TMEDA)]$, $[Ir(COD)_2BF_4]$, $[Rh(COD)_2BF_4]$ and a 1:1 mixture of $[PdCl_2(PhCN)_2]$ and $[Ni(CH_3)_2(TMEDA)]$ afforded the related metallodendrimers. The Rh(I) supported dendritic catalyst was used for the hydroformylation of 1-octene with a turnover number comparable to the monomer, and with the possibility to recover the catalyst by membrane separation techniques.[17] Hydrogenation of conjugated dienes using the $PdCl_2$ supported catalyst has been reported by Mizugaki et al.[18] The selectivity was found to be excellent with the conversion of cyclopentadiene into cyclopentene with a catalytic activity higher than the corresponding monomer $[PdCl_2\{PhN(CH_3PPH_2)_2\}]$.

Dendrimer supported transition metal complexes have also been used for the hydrogenation of olefins.[9,10,18,19-23] "Nanoreactors" were prepared by Crooks et al.[19] using dendrimers as templates for the inclusion of metal nanoclusters within the branched framework. Both independent studies led by Crooks[19] and Tomalia[20] were based on the internal functionalization of the PAMAM dendrimer and showed the formation of colloids by reduction of metal ions coordinated to the nitrogen ligands forming the dendrimer. Crooks assessed the catalytic efficiency of such a system with the hydrogenation of allylic alcohol into 1-propanol by Pd particles encapsulated in OH-terminated dendrimers in water.[21] The turnover frequency was found to be 218 $mol_{H2}/mol_{Pd}/h$ at 20 °C.

Rh(I) based catalysts are often used in hydrogenation of olefins due to the high affinity of Rh(I) towards H_2 oxidative addition.[6] Dendrimers have also served as supports for these Rh(I) transition metal complexes. For example, we reported the preparation of organophosphine dendrimers in which phosphine moieties were distributed throughout the backbone of the resulting macromolecule.[9] A divergent acid-base hydrolytic methodology was employed using $Me_2Si(NMe_2)_2$ and tri(hydroxypropyl)phosphine as reaction partners (Scheme 1). One strategy leading to the preparation of related metallodendrimers was to react the organophosphine dendrimers with $[Rh(COD)Cl]_2$ by a bridge splitting reaction. The other strategy directly started with the preparation of the complex $[Rh(COD)ClP\{(CH_2)_3OH\}_3]$ before building the dendritic backbone. In very few synthetic steps, dendrimer "**Rh$_{46}$**" was prepared, featuring easy accessible 46 Rh(I) centers. Hydrogenation of 1-decene was carried out at T = 25°C, with 20 bar H_2 pressure in benzene for a reaction time of 30 min in a 1:200 dendrimer-to-substrate ratio. The catalytic activities (turnover frequency = 400 $mol_{prod}/mol_{dendrimer}/h$) of the Rh(I) dendrimers (**Rh$_4$**, **Rh$_{10}$**, **Rh$_{22}$**, **Rh$_{46}$**) were found to be similar to that of the monomeric complex **Rh$_1$**. The catalytic efficiency of these metallodendrimers was assessed upon recycling **Rh$_{46}$** cascade under the same reaction conditions. A slight decrease of about 5% conversion was observed after the second cycle.

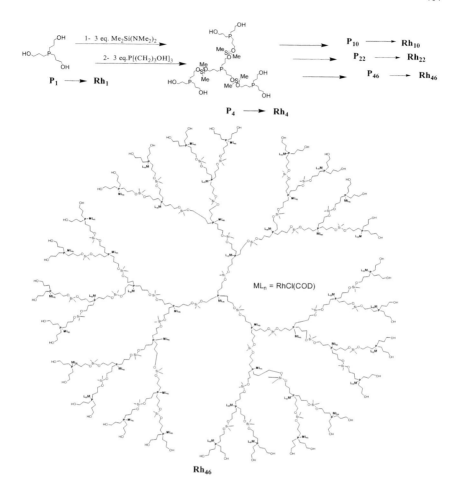

Scheme 1. Preparation of Rh(I) supported organophosphine dendrimers.[9]

The asymmetric hydrogenation by chiral centers located at the periphery of the dendrimer, has been studied by Togni et al.[22] Rh (I) complexes were prepared from ferrocenyldiphosphine ligand terminated dendrimers. The hydrogenation of dimethyl itaconate was performed with ee of 98%. Asymmetric hydrogenation by chiral centers located at the core[24, 25] and on the branches[26] of the dendrimer have also been reported. Rhodium catalyzed hydroformylation reaction of 1-octene was studied by Reek and van Leeuwen's group.[26] Functionalization of the diphenylphosphine terminated carbosilane dendrimers $Si\{(CH_2)_nSi(CH_3)_2(CH_2PPh_2)\}_4$ and $Si\{(CH_2)_nSi(CH_3)(CH_2PPh_2)_2\}_4$ (n = 2, 3) with Rh(I) catalysts afforded the related metallodendrimers. Selectivity was comparable to their monomer analogs and activities depended on the size and flexibility of the metallodendrimers. Diphenylphosphine functionalized dendrimer featuring a polyhedral oligomeric silsesquioxane core was reported by the Cole-Hamilton's group.[27] Functionalized with Rh(I) complexes, the resulting metallodendrimer catalyzed the hydroformylation of 1-octene and showed a higher selectivity than their small molecule analogs.

Hyperbranched Polymers as Supports for Catalysis

Though the use of dendrimers as supports for transition metal complexes is considered to bridge the gap between homogeneous and heterogenized catalysis, their synthesis can involve expensive, labour-intensive multistep procedures which limit the amount of available material. The preparation of hyperbranched polymers circumvents these synthetic difficulties but often results in materials with high polydispersity. Studies related to their use as supports for catalysts are seminal but remain promising. The van Koten's group has reported the preparation of Pd-supported hyperbranched polytriallylsilane.[28] The hyperbranched carbosilane support (HCS) was synthesized by polymerization of triallylsilane with platinum catalysis. Hydrosilylation of the poly(triallylsilane) with $HSiMe_2Cl$ was followed by the addition of 3,5-bis[[(dimethylamino)methyl]phenyl lithium (Li-NCN) to afford the coordinating ligand NCN-terminated hyperbranched polymer. After lithiation, the addition of the Pd(II) complex afforded the corresponding Pd(II) supported hyperbranched carbosilane. This new catalytic hyperbranched system was used for the catalytic aldol condensation of benzaldehyde and methyl isocyanate with $[Pd(OH_2)]OTf$ replacing the PdCl complexes. The turnover number per Pd site was shown to be slightly lower when using the hyperbranched catalyst compared with the monomer analog.

Search for catalysts featuring advantageous properties of both homogeneous and heterogeneous systems remains a topical area of research. The use of dendrimers in which transition metal catalytic centres can be distributed throughout the backbone or at the periphery, seems to offer a good compromise in this regard. In addition, the loading capacity increases exponentially as the dendrimers increase in size (generation number). Peripheral functionalization of 3,5-dihydroxybenzyl alcohol based dendrimers and hyperbranched polymers with Rh(I) catalysts is an attractive route to synthesize organometallic dendritic polymers,[10] since the supports are easily prepared in large amounts by following a simple acid-base hydrolytic methodology[30-35] using commercially available reagents 3,5-dihydroxybenzyl alcohol and $Me_2Si(NMe_2)_2$.[36] The catalytic activity of the resulting materials in hydrogenation of 1-decene was examined, and the influence of the distribution of the Rh(I) centers, at the periphery or throughout the backbone, was assessed by comparing the results obtained using the DHBA-based metallodendrimers with the ones containing Rh(I) sites distributed throughout the backbone reported previously by our research group.[9,10,36,37]

Results and Discussion

3,5-Dihydroxybenzyl Alcohol Based Dendrimers and Hyperbranched Polymers Functionalized with Rh(I) Metal Catalysts at Their Periphery

The 3,5-dihydroxybenzyl alcohol based dendrimers were prepared using an iterative sequential divergent acid base hydrolytic protocol involving the reaction of 3,5-dihydroxybenzyl alcohol and bis(dimethylamino)dimethylsilane.[10, 36] By controlling each reaction through a slow addition process, we were able to prepare hydroxyl group terminated dendrimers up to the fifth generation (Figure 1).

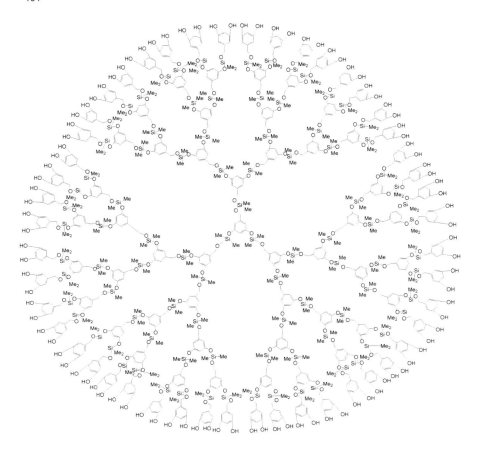

Figure 1. 3,5-dihydroxybenzylalcohol based 5th generation dendrimer.[36]

Because of their design as well as the flexibility of the acid-base hydrolysis synthetic methodology, the functionalization of the periphery with donor phosphine ligands that would coordinate with rhodium metal centers was easily achieved. A typical procedure used for this synthesis involved reacting 6 equivalents of Me$_2$Si(NMe$_2$)$_2$ with 1 equivalent of the first-generation dendrimer (**1**) in THF, followed by the addition of 6 equivalents of 3-hydroxypropyldiphenylphosphine (HO(CH$_2$)$_3$PPh$_2$, **2**) in THF.[10,36] The solution mixture was stirred at room temperature overnight, and the solvent was then removed under vacuum to yield the diphenylphosphine terminated first generation dendrimer (**3**) (Scheme 2).[10, 36] This procedure was repeated for dendrimer generations 2 and 3 using 12 and 24 equivalents of $HO(CH_2)_3PPh_2$ respectively resulting in diphenylphosphine-terminated second and third generation dendrimers (**4** and **5**). However, completion of the reaction of 3-hydroxypropyldiphenylphosphine with the peripheral aminosilane groups of the dendrimers required heat as well as additional time of reaction, as the steric congestion at the surface increased: 24 h at 55 °C to yield compound **4**, and 7 days at 55 °C to yield compound **5**.[36] This study focuses on functionalizing dendrimer generations 1-3 with phosphines followed by complexation with Rh(I) metal centers, as the steric congestion at the periphery of dendrimer generations 4 and 5 would prevent complete functionalization with the phosphine ligands.[10] $^{31}P\{^1H\}$NMR proved to be particularly useful to determine if the functionalization with the terminal phosphine units was complete. In the original mixture of aminosilane-terminated dendrimers and 3-hydroxypropyldiphenylphosphine, two peaks could be observed in the $^{31}P\{^1H\}$NMR spectra, one at -15.7 ppm for the $HO(CH_2)_3PPh_2$ reagent and the other at -16.0 ppm for the grafted $-O(CH_2)_3PPh_2$ moiety. Reaction completion was determined when the only peak at -16 ppm remained in the spectrum.[10, 36] MALDI-TOF mass spectroscopy was also very useful to characterize these metallodendrimers.

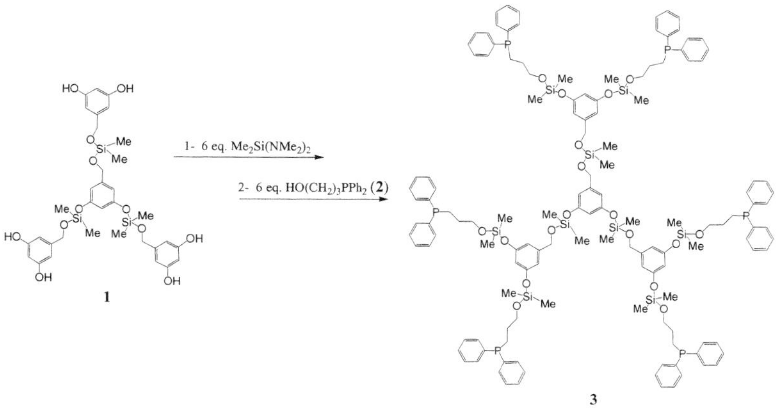

Scheme 2. Functionalization of first generation DHBA-based dendrimer with terminal 3-hydroxypropyldiphenylphosphine units.[10,36]

The related metallodendrimers were prepared by reacting phosphinated first-generation dendrimer (**3**) with [RhCl(COD)]$_2$ in a bridge-splitting reaction (Scheme 3).[10, 36] No residual free phosphine units at the periphery (**6**) were observed, as indicated by the complete disappearance of the singlet at -16.0 ppm and the appearance of a doublet at 27.3 ppm due to Rh-P coupling (J_{Rh-P} = 150 Hz) in the ^{31}P{^{1}H} NMR spectrum. Metallodendrimers **7** and **8** were prepared using similar procedure of reacting dendrimer generations 2 and 3 with 6 and 12 equivalents of the rhodium dimer respectively.[10, 36] Metallodendrimers **6-8** were recovered as yellow solids once the solvent was removed under vacuum.

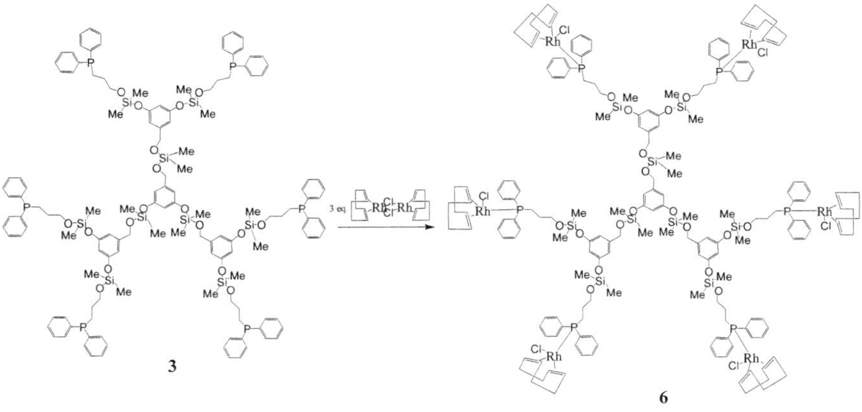

Scheme 3. Preparation of metallodendrimers by bridge splitting reaction of [Rh(COD)Cl]$_2$.[10,36]

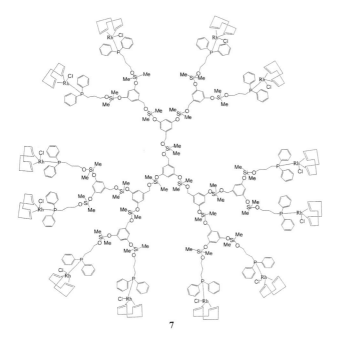

7

Figure 2

Functionalization of Hyperbranched Polymers

Preparation of DHBA-based hyperbranched polymers using a similar methodology as the one used for dendrimers,[36] led to the formation of a mixture of hyperbranched macromolecules. We were interested to know if their functionalization with Rh(I) complexes was possible, and how their results in hydrogenation of 1-decene would compare with that of metallodendrimers. The general synthetic strategy leading to the functionalization of the DHBA-based hyperbranched polymers[10, 36, 38] was the same as the one adopted for dendrimers. However, we had to take into consideration the fact that the synthesis of hyperbranched polymers leads to the formation different structural species. The preparation of hyperbranched polymers through 2 step-, 4 step- as well as 6 step-procedures resulted in multi-species mixtures with structures ranging from that of dendrimer generation 1 to the dendrimer generation 2. Hence, HP-2 step, HP-4 step and

HP-6 step were reacted with 6, 12 and 24 equivalents of $Me_2Si(NMe_2)_2$ and then 3-hydroxypropyldiphenylphosphine, respectively as in the case of their dendrimer analogs.[38] The number of moles of hyperbranched polymers used for the reaction was related to the initial molar quantity of DHBA involved during the first step of the preparation of the hyperbranched polymers. Reaction completion of the phosphine needed 3 days of stirring at 55°C for all of them. The reaction was monitored by $^{31}P\{^1H\}$ NMR spectroscopy. The completion of the reaction was assumed when a single peak was obtained at -16.0 ppm. The different mass fractions observed by MALDI-TOF mass spectroscopy were also indicative of diphenylphosphine units grafted on the hyperbranched backbone. For **[HP-2]-(PPh₂)**, masses ranged from 1315 amu with a probable structure depicted in compound **9**, to 3 475 amu (**10** as the possible structure) and including the diphenylphosphine-terminated dendrimer generation one (**3**) (2 530 amu).[38]

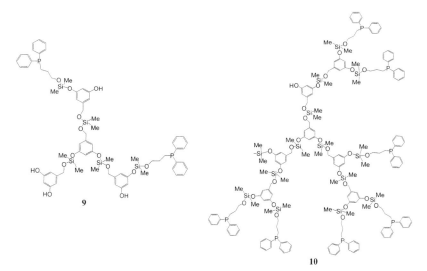

Figure 3

The main fractions observed for **[HP-4]-(PPh₂)** ranged from 1 315 (**9**), to 2 673 amu (**11**) also including compound **3**. The one for **[HP-6]-(PPh₂)** ranged from 1 315 (**9**) to 3 045 amu (**12**), again including compound **3**.[36, 38]

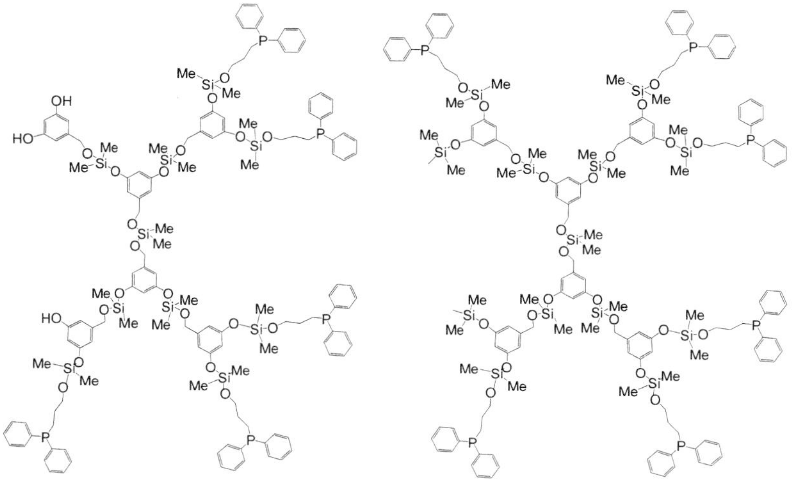

11   **12**

Figure 4

Reaction with the metal complexes was carried out using 3, 6 and 12 equivalents of Rh dimer with the phosphine functionalised HP-2 step, HP-4 step and HP-6 step respectively. Coordination to the rhodium center was monitored by ^{31}P{^{1}H} NMR with the appearance of a doublet at 27.3 ppm (J_{Rh-P} = 150 Hz).[38] In the case of the **[HP-2]-(Rh)**, MALDI fractions could be observed from 1 524 (**13**) to 5 289 (**14**) including Rh-functionalized first generation dendrimer (**6**). The mass fractions for **[HP-4]-(Rh)** from 1 848 (**15**) to 7 463 amu (**16**) and from 1162 (**17**) to 8 464 amu (**4**) for **[HP-6]-(Rh)**.[38]

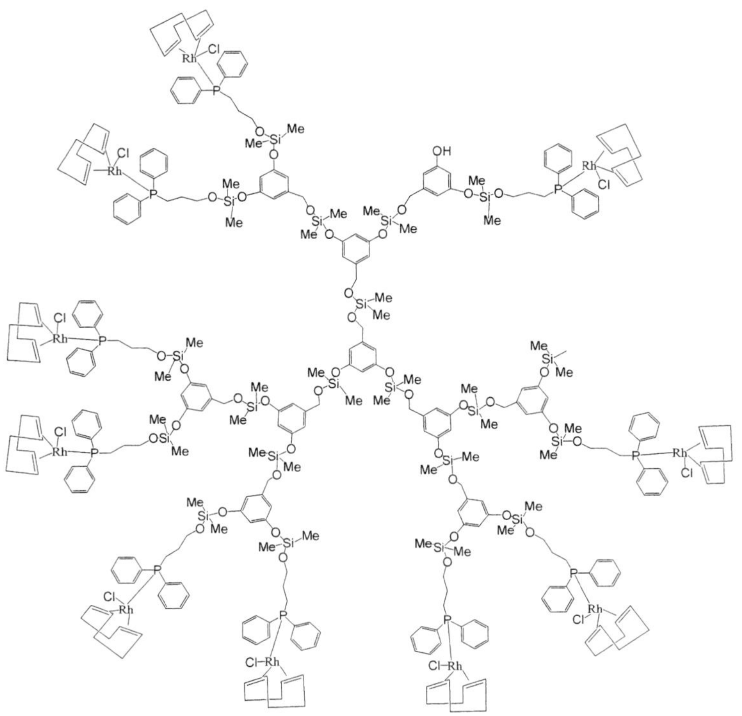

16

Figure 5

Hydrogenation of 1-Decene

Using Metallodendrimers Based on 3,5-Dihydroxybenzyl Alcohol Dendrimers[36,38]

The hydrogen pressure reactor was loaded with the organometallic dendrimer and decene (1:200 molar ratio) dissolved in benzene. Hydrogenation of decene was carried out at room temperature at 20 bars H_2 pressure for periods of 0.5 to 5 h, and these reaction conditions were used as standard throughout the study.[36,38] Distillation of the reaction mixture after catalysis afforded the recovery of the Rh(I) dendritic material as a residue.

The % conversion rates of 1-decene into decane are summarized in Table 1, and show dependence on two distinct factors, the time of reaction and the generation number.

Table 1. Conversion rates (in %) of 1-decene into decane.[10, 36, 38]

Dendrimer	Time of Reaction h					
	0.5	**1**	**2**		**5**	
			1st cycle	2nd cycle	1st cycle	2nd cycle
[G-1]-(Rh)$_6$	19	47	48	85	97	93
[G-2]-(Rh)$_{12}$	23	65	71	93	98	97
[G-3]-(Rh)$_{24}$	26	65	93	98	93	91

The catalytic activity increased with an increase of time of reaction. For example, with organometallic dendrimer of generation 1, only 19% of decene was found to be converted to decane after 0.5 h of the reaction which gradually increased to 97% at 5 h of the reaction time. The reactivity of organometallic dendrimer of generation 2 followed a similar pattern of an increase in activity with time. Dendrimer generation 3 showed a maximum conversion after 2 h of contact time.

If we compare the differences in catalytic activity between all three metallodendrimers for 0.5 h reaction time, results are similar (19, 23 and 26%), and do not show a significant increase when the generation number increases. A roughly similar trend can be observed for 1 h reaction time as [G-1]-(Rh)$_6$ converted 47% of decene into decane and both [G-2]-(Rh)$_{12}$ and [G-3]-(Rh)$_{24}$ showed 65% conversion. However, for a reaction time of 2 h, conversion rates of [G-1]-(Rh)$_6$, [G-2]-(Rh)$_{12}$ and [G-3]-(Rh)$_{24}$ are 48, 71 and 93% respectively.[10, 36, 38] Results for 5 h reaction time show maximum conversion rates for all three metallodendrimers. The results at lower time of contact suggest that the catalytic conversions of decene into decane are slower. It should be noted that increasing the generation number increases the number of Rh(I) centers at the periphery of the DHBA-based metallodendrimer. More decene is then likely to be converted into decane using higher generation number metallodendrimers.

By reloading the catalyst with fresh decene for another 2 h, we expected, at best, a conversion rate similar to that obtained at the end of the first cycle of 2 h. However, an increase in catalytic activity was observed for **[G-1]-(Rh)₆** (48% to 85%) and **[G-2]-(Rh)₁₂** (71% to 93%).[36,38] This increase in conversion rates suggests that the catalyst was activated in the first cycle of 2 h reaction.[38] This activation can be caused by having decene already bound to the Rh center. Displacement of COD by decene is one of the steps involved in the catalytic reaction, and having it taken care of in the first 2 h of the cycle will accelerate the reaction in the second cycle of 2 h. One can also consider the fact that the recycling reaction started with the active Rh(III) dihydride species. Results upon recycling of the catalysts after a period of 5 h did not lead to an increase in conversion rates that were already at their maximum after one cycle. They are actually comparable to that of the first cycle of 5 h.[36, 38]

The monomeric species $[HO(CH_2)_3PPh_2Rh(COD)Cl]$ **18** showed a catalytic conversion of decene to decane at 49% for 1 h reaction time. By determining the turnover frequency (TOF) of the metallodendrimers and that of the monomeric species (**18**), we could compare their catalytic activity with that of the Rh(I)-supported tri(alkyl)phosphine dendrimers in which the active catalytic centers are distributed throughout the backbone.[9] TOFs after 1h reaction time for monomer **18**, **[G-1]-(Rh)₆**, **[G-2]-(Rh)₁₂** and **[G-3]-(Rh)₂₄** were found to be 98, 94, 130 and 130 $mol_{prod}/mol_{cat}/h$ respectively.[10,36,38] However, TOFs per molecule of tri(alkyl)phosphine-based metallodendrimer were found to be 400 $mol_{prod}/mol_{cat}/h$ for **Rh₁**, **Rh₄**, **Rh₁₀**, **Rh₂₂** and **Rh₄₆** for a reaction time of 0.5 h.[9] By considering the number of catalytic sites supported on each species, TOFs of **18**, **[G-1]-(Rh)₆**, **[G-2]-(Rh)₁₂** and **[G-3]-(Rh)₂₄** can be compared to that of **Rh₁**, **Rh₄**, **Rh₁₀** and **Rh₂₂** respectively. The catalytic activities are lower for compound **18** as well as DHBA-based metallodendrimers (**[G-1]-(Rh)₆**, **[G-2]-(Rh)₁₂** and **[G-3]-(Rh)₂₄**), in which phosphine ligand features substituents of both aryl and alkyl nature. In addition, in the latter metallodendrimers, the catalytic sites are situated at the periphery. Such differences in catalytic efficiency can be explained by steric arguments as well as by difference of electron density at the metal center.

In DHBA-based and tri(alkyl)phosphine-based[9] metallodendrimers, distribution of the catalytic sites is different. For the tri(alkyl)phosphine-based dendrimers, the phosphine ligands are distributed throughout the entire backbone of the dendrimer whereas the phosphine ligands are only periphery situated in the case of the DHBA-based dendrimers. Coordination with Rh(I) complexes throughout the backbone reduces the steric congestion that can occur around the metal centers when they are located at the periphery only. Nevertheless, one could argue that distribution of Rh(I) centers throughout the backbone would lead to an increase in steric hindrance around them. However, it should be noted that there is inherent flexibility of the structural backbone of these dendrimers imparted by the alkyl branches. If one considers the schematic representation of **Rh₄₆** (Scheme 1), each Rh centre is far away from its neighbours, and there is less steric congestion around the metal center.

It is also known that steric congestion at the phosphine ligand can influence the rate of catalysis.[29] For example, it is known that triphenylphosphine ligand shields the metal.[29] The steric bulkiness of a phosphine ligand is generally measured by its cone angle.[39] Although undetermined, we can assume that the cone angle of the propyldiphenylphosphine ligand used in DHBA-based dendrimers will be in between 132° (for PEt₃) and 145° (for PPh₃).[29,39] It is safe to assume then that it is higher than tri(hydroxypropyl)phosphine. Higher cone angle will reflect higher congestion at the metal center, and the latter will reduce the accessibility of the substrate to the metal centre. This will lead to slower rates of olefin hydrogenation.[38]

Substituents at the phosphine ligand can also influence the conversion rates in hydrogenation of olefins.[5b,29] It is known that alkyl groups are more electron donating than aryl groups. Phosphines in which substituents are of alkyl nature are more electron rich, and when bound to a metal, will increase the electron density at the transition metal center. Kinetic studies led by Halpern[40] related to olefin hydrogenation have shown that the oxidative addition of H_2 occurred before binding of the olefin for Wilkinson-type Rh(I) catalysts.[5a,5b,29] If the Rh complexes, coordinated to tri(hydroxypropyl)-phosphine,[9] follow a similar reaction path, the oxidative addition should be enhanced in a more electron rich Rh(I) center. Because the Rh(I) centers are coordinated to hydroxylpropyldiphenylphosphine in the case of the DHBA-based metallodendrimers, the electron density at the Rh centre is expected to be lower than those in tri(alkyl)phosphine bound Rh centers. Hence, a slower addition of H_2 is expected. This is reflected in catalytic hydrogenation of decene to decane using model complexes (**18** and **Rh₁**).[36, 38] Tri(alkyl)phosphine monomeric species **Rh₁** shows a TOF of 400 $mol_{prod}/mol_{cat}/h$ whereas that for the monomeric species **18** is 98 $mol_{prod}/mol_{cat}/h$.[36] This argument can be extended to dendrimers as DHBA-based metallodendrimers show TOFs of 94, 130 and 130 $mol_{prod}/mol_{cat}/h$ for **[G-1]-(Rh)₆**, **[G-2]-(Rh)₁₂** and **[G-3]-(Rh)₂₄** respectively, that is much lower than tri(alkyl)phosphine metallodendrimers (400 $mol_{prod}/mol_{cat}/h$ for **Rh₄**, **Rh₁₀** and **Rh₂₂**).

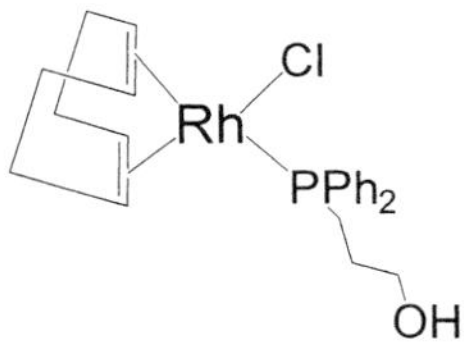

18   **Rh₁**

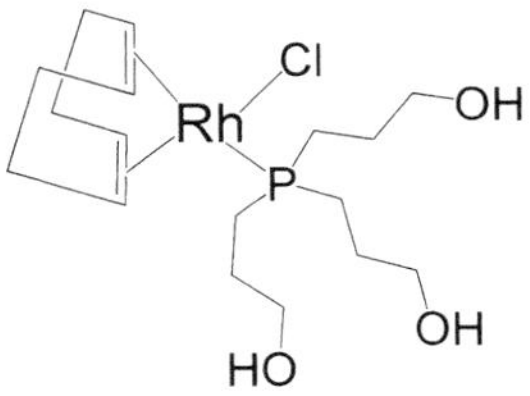

Figure 6

Using 3,5-Dihydroxybenzyl Alcohol Based Hyperbranched Polymers

The hydrogenation of 1-decene was also examined using Rh(I)-supported hyperbranched polymers. Reaction conditions were kept identical (RT, P = 20 bar H_2) to those used for the DHBA-based metallodendrimers. Conversion rates obtained from the hydrogenation reactions were intriguing (Table 2),[36, 38] and did not follow the trends observed in the case of metallodendrimers. **[HP-2]-(Rh)** and **[HP-4]-(Rh)** showed similar results (~ 80% conversion) for reaction times ranging from 0.5 to 2 h. Maximum conversion rates were obtained after 5 h reaction time as in the case of DHBA-based metallodendrimers (Table 1). Recycling of the catalyst for a 2 h reaction time did not increase the conversion rate as observed for the metallodendrimers. % conversion rates obtained using **[HP-6]-(Rh)** were comparable to those obtained using **[G-3]-(Rh)₂₄**. The catalytic activity increased gradually as the time of reaction was increased (26% after 0.5 h to 99% after 5 h). This difference in catalytic activity between **[HP-2]-(Rh)** and **[HP-4]-(Rh)**, and **[HP-6]-(Rh)** is intriguing, and some explanation may be obtained by analyzing the

MALDI-TOF MS data. Spectra obtained for **[HP-2]-(Rh)** and **[HP-4]-(Rh)** showed a multitude of peaks of different masses with similar peak heights. This suggests that both **[HP-2]-(Rh)** and **[HP-4]-(Rh)** are mixtures with a composition of organometallic macromolecules of similar ratio. MALDI-TOF spectrum for **[HP-6]-(Rh)** also showed a mixture of organometallic macromolecules, but revealed the dominance of one particular peak (compound **16**) for which peak height was much higher than the others. This may explain the differences in trend of the catalytic activity between **[HP-2]-(Rh)** and **[HP-4]-(Rh)**, and **[HP-6]-(Rh)**. The latter, if it could be considered as an almost single species organometallic hyperbranched polymer, would behave like a metallodendrimer. This might explain its catalytic activity, that is comparable to **[G-3]-(Rh)$_{24}$**. **[HP-2]-(Rh)** and **[HP-4]-(Rh)** polymers contain a mixture of organometallic hyperbranched species, and follow a different trend than **[HP-6]-(Rh)**.

Table 2. Conversion rates (in %) of 1-decene into decane for hyperbranched polymers.[36, 38]

Hyperbranched Polymer	Time of Reaction					
	h					
	0.5	**1**	**2**		**5**	
			1st cycle	2nd cycle	1st cycle	2nd cycle
[HP-2]-(Rh)	78	79	82	81	93	95
[HP-4]-(Rh)	86	83	84	80	100	99
[HP-6]-(Rh)	26	60	91	92	99	95

Conclusions

The search for the ideal catalytic system is a topic of continued high interest. Dendrimers are attractive supports since they can be tailored to host transition metal complexes at varied locations. The preparation of Rh(I) catalyst-supported dendrimers was rendered possible via the peripheral functionalization of the DHBA-based dendrimers with a tertiary phosphine by continuation of the divergent acid base hydrolytic procedure. Reaction of the dendrimers with $Me_2Si(NMe_2)_2$ resulted in peripheral aminosilane groups, needed to graft the 3-hydroxypropyldiphenylphosphine molecules. Reaction completion was found to require heat and supplemental time of reaction as the steric congestion increased in higher generations. Coordination of these phosphine ligands to transition metal centers was achieved by bridge splitting reaction with [Rh(COD)Cl]$_2$ resulting in dendrimers of generations 1-3 featuring 6, 12 and 24 Rh(I) catalytic sites respectively on their surface. A similar procedure was applied to functionalize the DHBA-based hyperbranched polymers with similar Rh(I) catalytic moieties.

The hydrogenation of 1-decene was studied using these DHBA-based organometallic dendritic materials under standard conditions (RT, P = 20 bar H$_2$) over different periods of reaction time ranging from 0.5 to 5 h. In the case of organometallic dendrimers, the catalytic activity was found to be dependent upon two distinct factors. An increase in time of reaction led to an increase in catalytic activity, as well as upon an increase in the generation number. Upon recycling the catalyst after a period of 2 h, a significant increase in % conversion rate was observed. Maximum conversion was observed for a reaction time of 5 h.

Comparing these results to previous study on the hydrogenation of 1-decene using dendrimers in which Rh(I) centers were distributed throughout the backbone,[9] indicated much slower conversion rates in periphery situated metallodendrimers. Steric arguments around the Rh centers, differences in bulkiness, and electron donating abilities between the tri(alkyl)phosphine[9] and the propyldiphenylphosphine can partly explain these differences in conversion rates. The nature of the substituents on the phosphine ligand is known to influence the sterics and electron density at the metal center. Oxidative addition of H$_2$ may be facilitated when the Rh center is more electron rich, as in the case of the tri(alkyl) substituted phosphine,[9] leading to higher catalytic activity than when the phosphine contains one alkyl group and two aryl groups. This may explain higher catalytic efficiency of tri(alkyl)phosphine-based metallodendrimers than DHBA-based metallodendrimers.

The hydrogenation of 1-decene using the Rh(I)-supported hyperbranched polymers led to good conversion rates with a reaction time as low as 0.5 h in the case of the **[HP-2]-(Rh)** and **[HP-4]-(Rh)**, whereas the catalytic activity increased gradually in the case of **[HP-6]-(Rh)**. This difference in behaviour can be related to the inherent composition of each hyperbranched polymer mixture. **[HP-2]-(Rh)** and **[HP-4]-(Rh)** are both mixtures of organometallic macromolecules of equal ratio, and **[HP-6]-(Rh)** mixture showed the presence of a single dominant species, from which the trend in catalytic activity can be compared to that of metallodendrimers.

Acknowledgements

We would like to thank NSERC (Canada) and FQRNT (Quebec, Canada) for financial support.

[1] R. van Heerbeek, P. C. J. Kamer, P. W. N. M. van Leeuwen, J. N. H. Reek, *Chem. Rev.* **2002**, *102*, 3717.
[2] A.K. Kakkar, *Chem. Rev.* **2002**, *102*, 3579.
[3] A. W. Kleij, R. A. Gossage, J. T. B. H. Jastrzebski, J. Boersma, S. J. E. Mulders, G. van Koten, *Angew. Chem., Int. Ed.* **2000**, *39*, 176.
[4] J. W. J. Knapen, A. W. van der Made, J. C. de Wilde, P. W. N. M. van Leeuwen, P. Wijkens, D. M. Grove, G. van Koten, *Nature* **1994**, *372*, 659.
[5] [5a] R. S. Dickson, "*Homogeneous Catalysis with Compounds of Rhodium and Iridium*", D. Reidel Publishing Company, Dordrecht 1985; [5b] P. A. Chaloner, M. A. Esteruelas, F. Joó, L. A. Oro, "*Homogeneous Hydrogenation*", Kluwer academic publishers, Norwell 1994; [5c] D. J. Law, R. G. Cavell, *J. Mol. Cat.* **1994**, *91*, 175.
[6] [6a] F. R. Hartley, "*Supported Metal Complexes*", D. Reidel Publishing Company, Dordrecht, 1985; [6b] C. Merckle, S. Haubrich, J. Blümel, *J. Organ. Chem.* **2001**, *627*, 44.
[7] F. Chardac, D. Astruc, *Chem. Rev.* **2001**, *101*, 2991.
[8] M. Dasgupta, M.B. Peori, A.K. Kakkar, *Coord. Chem. Rev.* **2002**, *233-234*, 223.
[9] M. Petrucci-Samija, V. Guillemette, M. Dasgupta, A. K. Kakkar, *J. Am. Chem. Soc.* **1999**, *121*, 1968.
[10] O. Bourrier, A.K. Kakkar, *J. Mater. Chem.* **2003**, *13*, 1306.
[11] N. J. Hovestad, J. L. Hoare, J. T. B. H. Jastrzebski, A. J. Canty, W. J. J. Smeets, A. L. Spek, G. van Koten, *Organometallics* **1999**, *18*, 268.
[12] S. B. Garber, J. S. Kingsbury, B. L. Gray, A. H. Hoveyda, *J. Am. Chem. Soc.* **2000**, *122*, 8168.
[13] A.-M. Caminade, J.-P. Majoral, *Top. Curr. Chem.* **1998**, *197*, 79.
[14] M. Bardaji, M. Kustos, A.-M. Caminade, J.-P. Majoral, B. Chaudret, *Organometallics* **1997**, *16*, 403.
[15] M. Bardaji, A.-M. Caminade, J.-P. Majoral, B. Chaudret, *Organometallics* **1997**, *16*, 3489.
[16] V. Maraval, A.-M. Caminade, J.-P. Majoral, *Organometallics* **2000**, *19*, 4025.
[17] M. T. Reetz, G. Lohmer, R. Schwickardi, *Angew. Chem., Int. Ed.* **1997**, *36*, 1526.
[18] T. Mizugaki, M. Ooe, K. Ebitani, K. Kaneda, *J. Mol. Catal. A* **1999**, *145*, 329.
[19] M. Zhao, L. Sun, R. M. Crooks, *J. Am. Chem. Soc.* **1998**, *120*, 4877.
[20] L. Balogh, D. A. Tomalia, *J. Am. Chem. Soc.* **1998**, *120*, 7355.
[21] M. Zhao, R. M. Crooks, *Angew. Chem., Int. Ed.* **1999**, *38*, 364.
[22] C. Köllner, B. Pugin, A. Togni, *J. Am. Chem. Soc.* **1998**, *120*, 10274.
[23] Q.-H. Fan, Y.-M. Chen, D.-Z. Jiang, F. Xi, ASC. Chan, *Chem. Comm.* **2000**, 789.
[24] S. Yamago, M. Furukawa, A. Azuma, J.-I. Yoshida, *Tetrahedron Lett.* **1998**, *39*, 3783.
[25] [25a] H. J. Brunner, *J. Organomet. Chem.* **1995**, *500*, 39; [25b] H. J. Brunner, S. Altmann, *Chem. Ber.* **1994**, 127, 2285. [25c] H. Brunner, M. Janura, S. Stefaniak, *Synthesis* **1998**, 1742.
[26] D. De Groot, P. G. Emmerink, C. Coucke, J. N. H. Reek, P. C. J. Kamer, P. W. N. M. van Leeuwen, *Inorg. Chem. Comm.* **2000**, *3*, 711.

[27] [27a] L. Ropartz, R. E. Morris, D. F. Foster, D. J. Cole-Hamilton, *Chem. Comm.* **2001**, 361. [27b] L. Ropartz, R. E. Morris, G. P. Schwartz, D. F. Foster, D. J. Cole-Hamilton, *Inorg. Chem. Comm.* **2000**, *3*, 714.

[28] C. Schlenk, A. W. Kleij, H. Frey, G. van Koten, *Angew. Chem., Int. Ed.* **2000**, *39*, 3445.

[29] J. P. Collman, L. S. Hegedus, J. R. Norton, R. G. Fincke, *"Principles and Applications of Organotransition Metal Chemistry"*, University Science books, Mill Valley 1987.

[30] C.M. Yam, A.J. Dickie, A.K. Kakkar, *Langmuir*, **2002**, *18*, 8481.

[31] A.J. Dickie, A.K. Kakkar, M.A. Whitehead, *Langmuir*, **2002**, *18*, 5657.

[32] H. Jiang and A.K. Kakkar, *J. Am. Chem. Soc.* **1999**, *121*, 3657.

[33] M.G.L. Petrucci, D. Fenwick, A.K. Kakkar, *J. Mol. Catal. A: Chemical*, **1999**, *146*, 309.

[34] F. Chaumel, H. Jiang, A.K. Kakkar, *Chem. Mater.* **2001**, *13*, 3389.

[35] O. Bourrier, J. Butlin, R. Hourani, A.K. Kakkar, *Inorg. Chim. Acta*, Special issue, Manuscript in preparation.

[36] O. Bourrier, Ph.D. Thesis 2004, McGill University.

[37] M. Brad Peori, A.K. Kakkar, *Organometallics*, **2002**, *21*, 3860.

[38] O. Bourrier, A.K. Kakkar, unpublished results.

[39] C. A. Tolman, *Chem. Rev.* **1977**, *77*, 313.

[40] J. Halpern, *Inorg. Chim. Acta* **1981**, *50*, 11.

Macromol. Symp. **2004**, *209*, 119-131

Ring-Opening Polymerisation of Coordination Rings and Cages

Stuart L. James

School of Chemistry, Queens University Belfast, David Keir Building, Stranmillis Road, Belfast, Northern Ireland, BT9 5AG, UK

E-mail: s.james@qub.ac.uk

Summary: Despite the great interest in crystalline coordination polymers (sometimes called metal organic frameworks) surprisingly little is known about how they form. However, there has recently been some attention given to characterising their solution-based precursors. Of particular interest is the formal ring-opening polymerisation (ROP) relationship between the structures of some precursors and polymers, and the actual observation of ring-opened oligomers by solution state NMR spectroscopy. These points are highlighted and discussed along with related aspects of isomerism in discrete and polymeric coordination structures, and polymer-polymer interconversion.

Keywords: coordination complex; coordination polymer; metal organic framework; MOF; ring-opening polymerisation; ROP

Introduction

The study of coordination polymers has increased enormously over the past ten years.[1] Despite this widespread activity, surprisingly little has been reported on how these materials might form. Since most syntheses involve crystallisation of the polymer from solution, it is of interest in this regard to identify solution-based precursors to the polymers. In this article the recent papers for which this has been an important aspect are summarised and discussed. Also, the related aspects of 'supramolecular isomerism' (discrete and polymeric isomers prepared from the same metal and ligand blocks), and polymer-to-polymer interconversion are highlighted.

 DOI: 10.1002/masy.200450508

Review and Discussion

i) Apparent ROP of Solution-based Rings and Cages upon Crystallisation

In 2001 we reported that when crystals of the dinuclear triply-bridged complex $[Ag_2(dppa)_3(OTf)_2]$ (dppa = *bis*(diphenylphosphino)acetylene), which is also the main species present in solutions, were left in their supernatant for 16 weeks, they transformed by redissolution and recrystallisation into a polymeric form $[Ag_2(dppa)_3(OTf)_2]_\infty$.[2] The polymer consists of dinculear rings, in which the two silver ions are bridged by two diphosphines, and a third diphosphine links the rings into a one-dimensional chain (Figure 1).

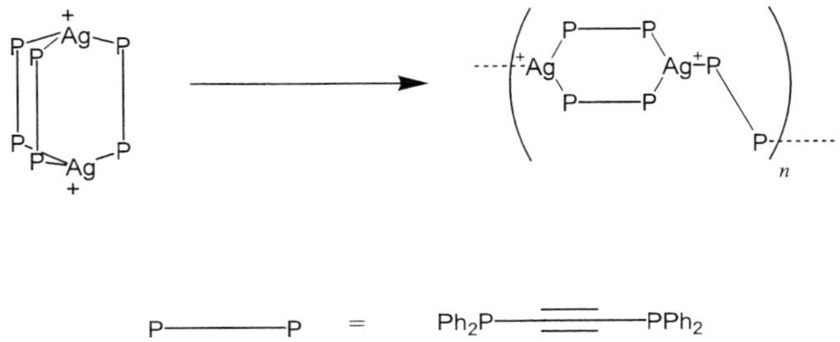

Figure 1. Apparent ring-opening polymerisation of triply-bridge disilver complexes $[Ag_2(dppa)_3(anion)_2]$ to give a linear polymer.

It was intriguing that the discrete and polymeric complexes were formally related to each other by ring-opening polymerisation (ROP). Although ROP is well known in main group chemistry and for strained metallocenes it had not to our knowledge been reported previously for coordination cages. A key difference is that for the ROP of kinetically-stabilised main group rings there is direct and irreversible transformation to the polymer, whereas in these coordination complexes the formation of M-L bonds is reversible. Therefore, potentially there are many equilibira to be considered in the formation of a coordination polymer, for example those shown in Figure 2, which provide indirect routes to the observed polymer. We postulated that a direct ROP of the precursor to the polymer *could*, nevertheless, be the actual mechanism by which the polymer formed. Related work with the slightly more bulky diphosphine ligand dppet, *trans*-1,2-*bis*(diphenylphosphino)ethylene, suggested that the discrete dinuclear complexes were less stable than the dppa analogues in solution, since their NMR spectra were only resolved at low temperature, indicating that greater Ag-P dissociation occured than in the analogous dppa complexes. Consistent with this observation, the discrete complexes could not be isolated by crystallisation, and instead the polymeric form $[Ag_2(dppet)_3(SbF_6)_2]_\infty$ was obtained directly. ROP therefore appeared to be favoured by prolonged standing of crystals of the precursor in their supernatant and by steric destabilisaing of the precursor.

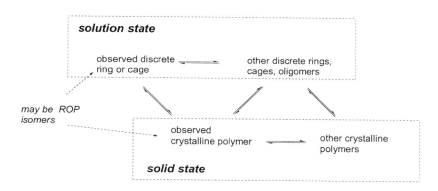

Figure 2. Some of the many potential equilibria involved in forming the observed polymeric product from a dynamic coordination system.

Simultaneously, Puddephatt and coworkers reported related coordination polymers based on gold. As well as one polymer, $[Au_2(dppet)_3](CF_3CO_2)_2$,[3] which was topologically the same as the silver-dppa and -dppet polymers we described, a chickenwire sheet polymer of formula $[Au_2(dppb)_3](AuCl_2)_2$ (dppb = 1,4-bis(diphenylphosphino)butane) was also described (Figure 3). The latter was formally related to a hypothesised solution-based triply-bridged precursor complex $[Au_2(diphosphine)_3]^{2+}$ by a double ROP process, *i.e.* by the opening of two rings in the precursor. Subsequently this group reported further on the ROP of coordination rings. [4-7] In particular, coordination rings containing two gold centres, linked by one bridging dppet diphosphine and one bridging diamine ligand, were proposed to undergo Au-N bond dissociation as a first step to oligomerisation, and ultimately to form linear polymers in which Au(I) centres were bridged alternately by the diphosphine and diamine ligands (Figure 4).[4] Such a polymer was in fact obtained with the alternative more flexible diphosphine ligand dppe, $Ph_2PCH_2CH_2PPh_2$. Variable temperature ^{31}P NMR spectroscopy was used to probe the solution structures of the complexes formed by each diphosphine. With the rigid ligand dppet, a single resonance at all temperatures was ascribed to the dinuclear ring as characterised by X-ray crystallography. However, with the more flexible linker dppe, more complex behaviour was seen which was both concentration- and temperature-dependent. Peaks resolved at low temperature were ascribed to the formation of low-molecular weight ring-opened oligomeric forms (Figure 4). The proportion of oligomers increased with the concentration. Further examples of solution-based ring structures, which transformed to polymers on crystallisation, were subsequently reported. These include mixed-ligand rings composed of the bridging diphosphine dppet and dithiols.[5] Also, use of a *bis*(pyridyl)-derived ligand, containing potentially hydrogen bonding amide functions, gave a tetranuclear ring structure when the other bridging ligand present was dppp, $Ph_2PCH_2CH_2CH_2PPh_2$, but a linear polymer when dppb, $Ph_2PCH_2CH_2CH_2CH_2PPh_2$, was used.[6] The low temperature ^{31}P NMR spectrum of the ring complex showed additional peaks which were assigned to the ring-opened form, in which anions replaced the N-donor ligands at the Au(I) centres. Other polymers based on Ag(I) and diphosphines, with a variety of topologies were also prepared and their relationships to discrete isomers discussed.[7]

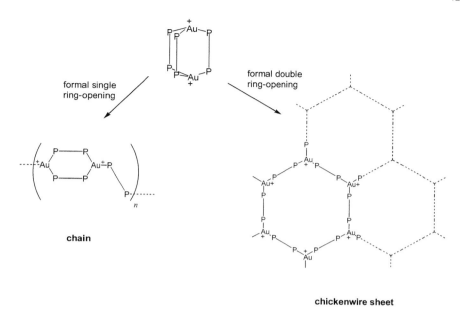

Figure 3. Formal single- and double-ring-opening relationships between a discrete triply-bridged complex and chain and chickenwire sheet polymers.

124

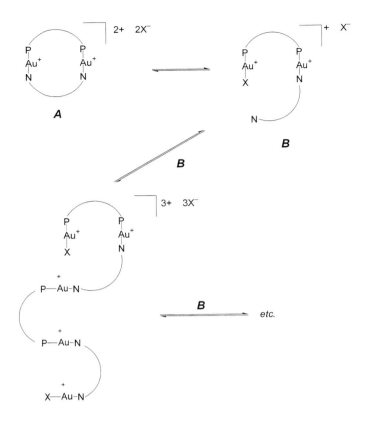

Figure 4. Proposed ROP pathway to polymers based on bridging diphosphines and bridging diamines (adapted from ref. 4, gold-gold contacts have been omitted for clarity).

Whereas the above work has involved phosphines as bridging ligands, Chung *et al.* have reported on apparent ROP of metallamacrocycles with purely N-based ligands.[8] Flexible ligands based on 1,3-*bis*(2-pyridyl)propane gave disilver macrocycles on crystallisation with silver salts. However, with 1,3-*bis*(2-pyridyl)-2-phenylpropane a linear polymer was obtained on crystallisation (Figure 5). This polymer was found to be slightly soluble in water, and ^{1}H NMR spectroscopy and solution molecular weight measurement showed that the dominant species present in solution was a dinuclear macrocycle similar to that isolated for the other ligands. The authors concluded that the polymer crystallised from a solution in which the predominant species was the discrete macrocycle.

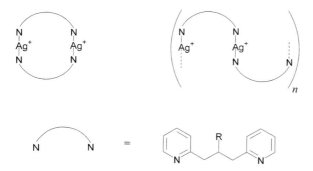

Figure 5. Discrete disilver and chain polymer structures obtained with bridging ligands based on 1,3-di(2-pyridyl)propane (anions have been omitted for clarity, in the polymeric structure they bridge between the chains).

As an interesting contrast to the study of crystalline coordination polymers, a non-crystalline system has been studied by Hill et al.,[9a] Gray et al.[9b] and Sijbesma et al.[9c] Reaction between PdCl$_2$ and the long chain diphosphine 1,12-bis(diphenylphosphino)dodecane in chloroform solvent gives mixtures of cyclic oligomers or linear polymer depending on the concentration (Figure 6). Diffusion-ordered ^{31}P NMR spectroscopy was used to identify the various cyclic oligomeric and linear polymeric species. There is a critical concentration below which only cycles are present, and above which the concentration of cycles stays constant and the amount of linear polymer increases. Slow evaporation of chloroform solutions gave viscous material containing both cyclic oligomers and linear polymer of up to 80 repeat units. However, melt polymerisation gave polymer of up to 500 repeat units.

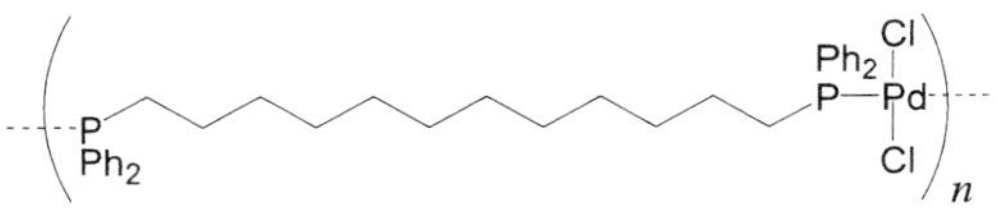

$n_{max} > 500$

Figure 6. Cyclic coordination oligomers and high molecular weight linear polymers described by Sijbesma *et al.*[9c]

ii) Ring-opening Isomerism

The issue of ROP of coordination rings and cages is closely related to what has been called 'supramolecular isomerisim',[10] or 'supramolecular ring-opening isomerism'.[11] In these cases, the emphasis has not been on the actual mechanisms by which coordination polymers form, but instead to highlight isomeric relationships between discrete and polymeric coordination compounds. It therefore provides a way of systematising some structures. For example, the hexagonal Cu_6 ring obtained by Zaworotko *et al.*[10] is a finite isomer of the corresponding zig-zag polymer, which was also obtained by these workers under slightly different reaction conditions (Figure 7). Zur Loye et al. reported complexation of HgX_2 salts with the semi-rigid ditopic ligand 1,3-*bis*(benzimidazol-1-ylmethyl)-2,4,6-trimethylbenzene (bbimms) (Figure 8).[11] A discrete binuclear ring of formula $Hg_2L_4L_2$ was isolated, which consisted of a metallacycle with two bridging organic ligands and four termial I^- anions. Its polymeric counterpart, the formal ROP isomer, having a chain structure of HgI_2 centres linked by bridging organic ligands was isolated under slightly different conditions. Solution-based structures were not characterised, but attention was drawn to the 'supramolecular ring-opening isomerism' between the two complexes. The potential usefulness of this concept in predicting discrete and polymeric coordination structures was emphasised. Other examples of discrete and polymeric complexes based on bridging phosphines, which have formal mutual ROP-relationships have been described by Catalano et al.[12] in gold(I) polymers containing 3,6-bis(diphenylphosphino)pyridazine, and Wong et al. in Ag(I) complexes based on $Ph_2PC_2H_4CONHCH_2CH_2NHCOCH_2CH_2PPh_2$.[13]

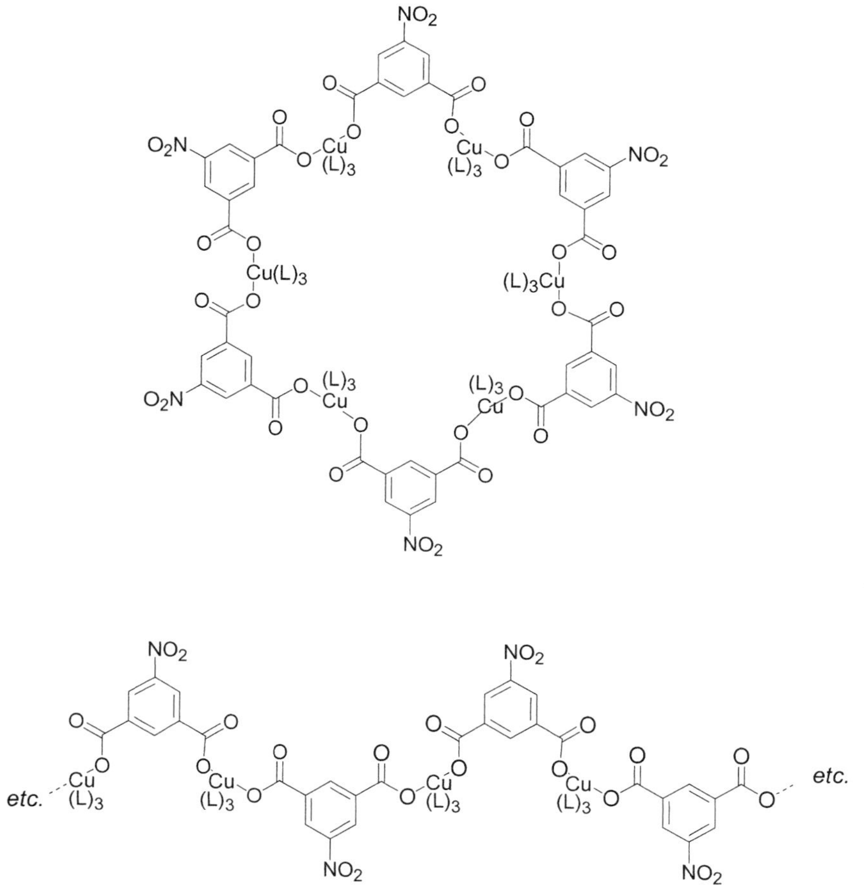

Figure 7. Discrete hexagonal and zig-zag polymeric 'supramolecular isomers' reported by Zaworotko et al.[10]

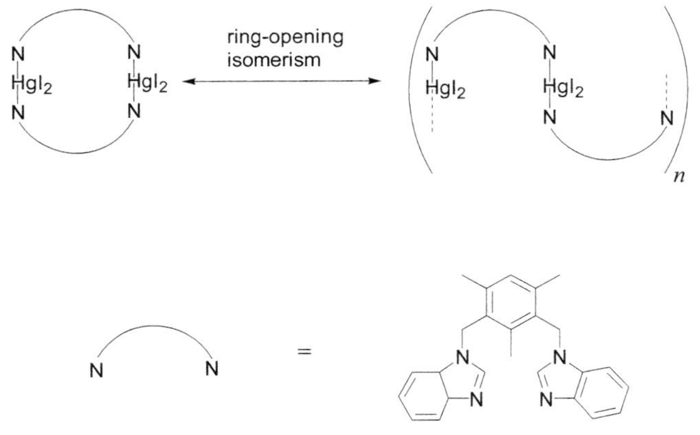

Figure 8. The ring-opening isomerism between a discrete dinuclear ring and linear polymer described by Zur Loye et al. (adapted from ref. 11).

iii) Solvent-mediated and Solid-state Interconversion of Isomeric Polymers

As well as isomerism between discrete and polymeric structures, isomerism can occur between polymers, and these polymers can sometimes also be interconverted. As well as being interesting from a structural point of view, these examples also demonstrate the high lability of the coordination systems which are often used to synthesise coordination polymers. They in fact provide examples of the various ‚indirect' equilibria shown in Figure 2. A recent example of solution-mediated polymer-to-polymer interconversion was described by Fromm et al.,[14] who obtained two coordination polymers by reacting CuCl with a *bis*(isonicotinic acid) derived ligand (Figure 9). Red crystals of one polymer, [CuCl(L)].0.5THF were obtained from an acetonitrile-THF solution of the ligand and CuCl on standing for a few hours. The structure consists of *spiro* chains of four-membered rings, defined by Cu and bridging Cl centres, alternating with larger 30-membered rings defined by Cu centres and the bridging organic ligands (Figure 9). The large rings leave sufficient space for included solvent molecules, which were modelled as disordered THF. Interestingly, when only acetonitrile was used as the crystallising solvent, red crystals were again obtained on standing for a few hours. These were then observed to redissolve and orange crystals were then obtained after a period of two months. The chain structure of the orange polymer did not contain rings, the Cu(I) centres being simply linked by bridging bidentate organic ligands, and having terminal Cl⁻ anions. It was postulated that the red crystals initially obtained from acetonitrile had the same *spiro* structure as those obtained from acetonitrile-THF. The two different chains are formally related to each other by the breaking of one Cu-N and one Cu-Cl bond, and the formation of one Cu-N bond per Cu centre. It was noted that the transformation from red to organge crystals stopped if the supernatatant was removed, showing the process to be solvent-mediated.

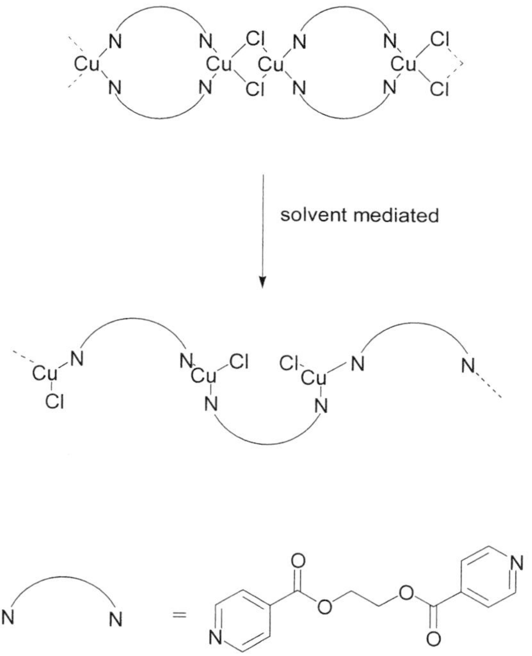

Figure 9. The solvent-mediated transformation of isomeric coordination polymers reported by Fromm et al.[14]

An example of solid-to-solid polymer interconversion was reported by Chung et al.[15] Reaction of $Co(NCS)_2$ with 1,2-dipyridylpropane gave, initially needle crystals of one polymer, followed by block crystals of another. Both had the formula $[Co(L)_2(NCS)_2]_n$ but had contrasting topologies. The first-formed consisted of interpenetrated rhombic grids with about 49% of the volume occupied by solvent molecules. The other consisted of two independent but mutually interpenetrated 3-D frameworks, both of which inteprenetrated with a 2-D grid. This polymer also had spaces occupied by solvent molecules, but accounting for only 18% of the total volume. On the basis of its greater available volume, the authors postulated that the first structure was metastable with regard to the second. They found that when crystals of the less dense polymer were removed from their supernatant, they lost solvent to become opaque, and XRPD showed that during this drying out the networks indeed transformed to the second, denser structure.

Conclusion

There are now several examples of discrete soluble species which have been found to give coordination polymers on crystallisation, and for which there is a formal ROP relationship between the precursor and the polymer.[2, 4, 5, 6, 8, 9] Opportunities and challenges in further establishing and exploiting this 'coordination ROP' include *i.* examination of precursor-polymer relationships in more cases, *ii.* evaluating the importance of potential indirect pathways to the polymer, as indicated in Figure 2, *iii.* detailed examination of the crystallisation process (potentially to get evidence of ROP at the crystal surface), *iv.* examination of how variables such as counter ion and solvent nucleophilicity, steric destabilisation and ring strain affect the ring-opening of precursors, and *v.* the deliberate design or prediction of network topologies based on the ROP principle.

[1] [1a] S. L. James, *Chem. Soc. Rev.* **2003**, *32*, 276; [1b] C. Janiak *J. Chem. Soc., Dalton Trans.* **2003**, 2781; [1c] B. Moulton, M. J. Zaworotko, *Curr. Opin. Solid State Mater. Sci.* **2002**, *6*, 117; [1d] B. Moulton, M. J. Zaworotko, *Chem. Rev.* **2001**, *101*, 1629; [1e] M. Eddaoudi, D. B. Moler, H. Li, B. Chen, T. M. Reineke, M. O'Keeffe, O.M. Yaghi, *Acc. Chem. Res.* **2001**, *34*, 319.

[2] E. Lozano, M. Niewenhuyzen, S. L. James, *Chem. Eur. J.* **2001**, *12*, 1644.

[3] M.C. Brandys, R.J. Puddephatt, *J. Am. Chem. Soc.* **2001**, *123*, 4839.

[4]   Z. Q. Qin, M. C. Jennings, R. J. Puddephatt, *Chem. Eur. J.,* **2002**, *8*, 735.

[5]   W.J. Hunks, M.C. Jennings and R.J. Puddephatt *Chem. Commun.* **2002**, 1834.

[6]   T. J. Burchell, D. J. Eisler, M. C. Jennings, R. J. Puddephatt, *Chem. Commun.* **2003**, 2228.

[7]   M. C. Brandys, R. J. Puddephatt, *J. Am. Chem. Soc.* **2002**, *124*, 3946.

[8]   D. M. Shin, I. S. Lee, Y. A. Lee, Y. K. Chung, *Inorg. Chem.* **2003**, *42*, 2977.

[9]   [9a] W. E. Hill, C. McAuliffe, I. E. Niven, R. V. Parish, *Inorg. Chim. Acta* **1980**, *38*, 273; [9b] D. C. Smith, G. M. Gray, *J. Chem. Soc., Dalton Trans.* **2000**, 677. [9c] J. M. J. Paulusse, R. P. Sijbesma, *Chem. Commun.* **2003**, 1494.

[10] H. Abourahma, B. Moulton, V. Kravtsov, M. J. Zaworotko *J. Am. Chem. Soc.* **2002**, *124*, 9990.

[11] C. Y. Su, A. M. Goforth, M. D. Smith, H. C. zur Loye, *Inorg. Chem.* **2003**, *42*, 5685.

[12] V. J. Catalano, M. A. Malwitz, S. J. Horner, J. Vasquez, *Inorg. Chem.* **2003**, *42*, 2141.

[13] Y. Li, K. F. Yung, H. S. Chan, W. T. Wong, W. K. Wong, M. C. Tse, *Inorg. Chem. Commun.* **2003**, *6*, 1315.

[14] A. Y. Robin, K. M. Fromm, H. Goesmann, G. Bernadinelli, *CrystEngComm*, **2003**, *5*, 405.

[15] D. M. Shin, I. S. Lee, D. Cho, Y. K. Chung, *Inorg. Chem.* **2003**, *42*, 7722.

Macromol. Symp. **2004**, *209*, 133-139

Metal-Containing Conjugated Oligo- and Polythiophenes

*Carolyn Moorlag, Olivier Clot, Yongbao Zhu, Michael O. Wolf**

Department of Chemistry, University of British Columbia, Vancouver, BC, Canada, V6T 1Z1

E-mail: mwolf@chem.ubc.ca

Summary: Conjugated polymers containing metal groups are an important class of new materials that may find application as sensors, non-linear optical materials and in molecular electronics. Polythiophenes are an extensively studied class of conjugated polymers, that can be easily modified synthetically, and may be prepared using a variety of methods. Metal groups pendant to polythiophene backbones modulate the electronic, optical and redox properties of the conjugated backbone and can introduce novel structural motifs to these materials. Functionalization of oligo- and polythiophenes by Fe, Pd and Ru containing groups is discussed, along with the properties of some of these materials.

Keywords: conductivity; conjugated polymers; electropolymerization; metal-containing polymers; optical properties; redox properties

Introduction

Polymers containing transition-metal complexes attached to or directly in a π-conjugated backbone are a promising class of modern materials. These hybrids of π-conjugated organic and transition-metal containing polymers allow the electronic properties of the conjugated backbone to be combined with the electronic, optical and catalytic properties of metal complexes. π-Conjugated organic polymers, including polyacetylene, polypyrrole, and polythiophene, as well as oligomers of these materials, have been studied extensively.[1] These materials are endowed with many important properties, such as non-linear optical properties, electronic conductivity and luminescence, and have been proposed for use in many applications including chemical sensors, electroluminescent devices, electrocatalysis, batteries, smart windows and memory devices.[1, 2]

Synthetic approaches to metal-containing conjugated polymers include condensation routes,[3,4] ring-opening metathesis polymerization,[5,6,7] and electropolymerization.[8]

DOI: 10.1002/masy.200450509

Electropolymerization was first used to prepare organic polymers such as polythiophene and polyaniline,[9] and results in deposition of insoluble electrogenerated material directly onto the electrode surface as a thin film. This allows oxidative or reductive doping to the conductive form to be achieved *in situ* electrochemically, allowing these polymers to be easily switched from the conductive to insulating state. Our group has focussed on preparing materials based on oligo- and polythiophenes due to their relative ease of synthesis relative to other conjugated materials.

Many different structural motifs have been incorporated into electropolymerized polythiophenes containing metals, including pendant Schiff base complexes[10] (**1**) or metal clusters[11] (**2**), and the properties of films of these materials explored.

1 M = Co, Zn, Cu   **2**

We have shown that electropolymerization of the ferrocene-containing monomer **4** and **5** (prepared as shown in Figure 1) results in films that show electroactivity both due to the $Fe^{II/III}$ couple and oxidation of the thiophene groups.[12]

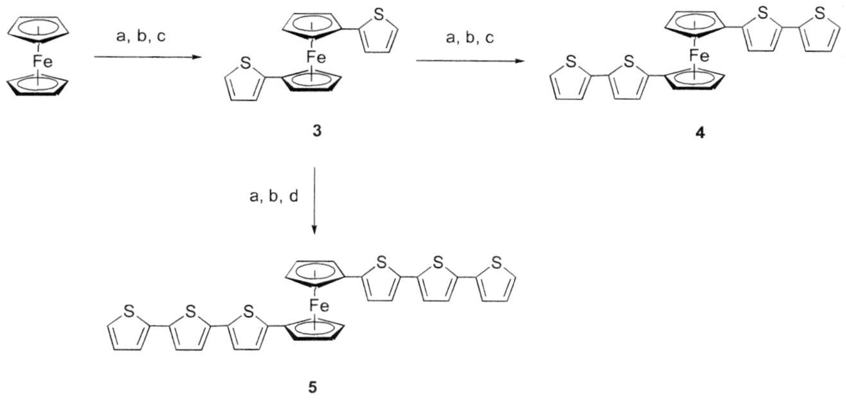

Figure 1. Synthesis of monomers **4** and **5**. Reagants: (a) BuLi, TMEDA, THF/hexanes. (b) ZnCl$_2$, THF. (c) 2-bromothiophene, Pd(PPh$_3$)$_4$, THF. (d) 5-bromo-2,2'-bithiophene, Pd(PPh$_3$)$_4$, THF.

Upon oxidation of a film of poly-**4** to a potential at which the ferrocenyl groups are oxidized, changes are observed in the UV-vis spectra of the films. Two bands are observed, at 495 and 1 395 nm. The broad, higher energy band is predominantly due to the π-π^* transition with contributions from Cp→Fe^{III} LMCT also possible; and the band with λ_{max} at 1 395 nm is assigned as a charge transfer band from the oligothiophene group to the Fe^{III}. Further oxidation of a poly-**4** film to 1.5 – 1.7 V vs. SCE results in the appearance of a very broad absorption between 400 and 1 600 nm. The appearance and positions of these bands suggest that they arise due to transitions to intergap states in the oxidized polymers, similar to transitions which appear upon oxidation of polythiophene.[13]

Pd-containing polymers may be prepared by electropolymerization of monomers **7-9**.[14, 15] These monomers are prepared by reaction of PdCl$_2$ with the phosphine substituted terthiophene **6** as shown in Figure 2.

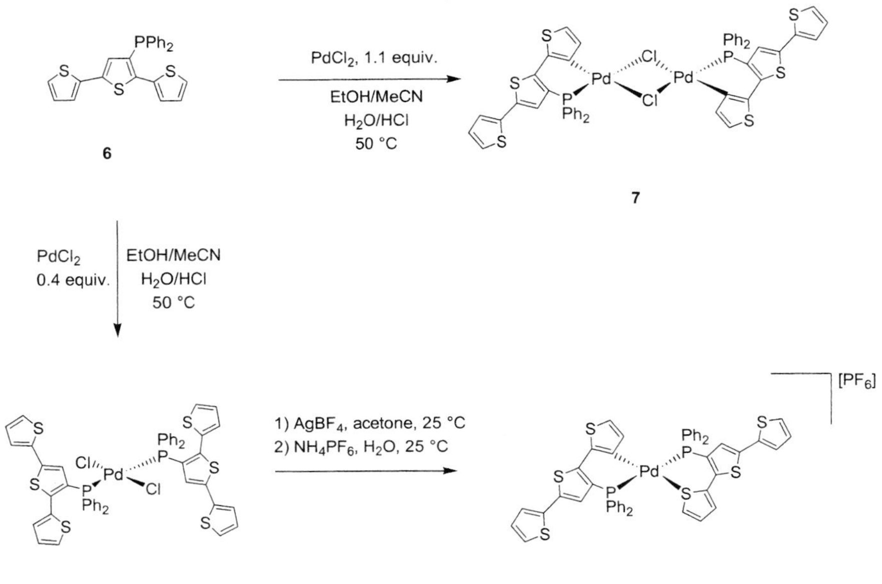

Figure 2. Synthesis of Pd containing monomers **7-9**.

Pd complexes **7-9** all electropolymerize when solutions of these monomers are scanned repeatedly over their whole oxidation range. Poly-**7** forms as a purple film that is conductive (10^{-3} S cm^{-1}) when oxidized. Poly-**8** forms as a red film and the voltammogram of the film shows a broad redox feature between +0.7 and +1.6 V, similar to that observed for poly-**7**. The maximum conductivity of oxidized poly-**8** is 3×10^{-4} S cm^{-1}, determined *in situ* by deposition of the film on interdigitated Pt microelectrodes. Similarly, electropolymerization of **9** (Figure 3) gave red films (reduced) with a conductivity of 10^{-4} S cm^{-1} when oxidized.

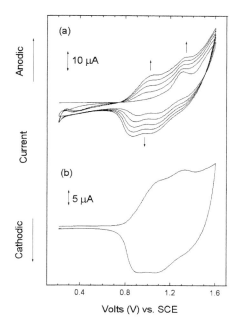

Figure 3. Cyclic voltammetry of **9** in CH₂Cl₂ containing 0.1 M [(*n*-Bu)₄NPF₆], scan rate = 200 mV/s. (a) Multiple scans from +0.20 to +1.60V. (b) Scan of a film of poly-**9** on a gold electrode.

The spectrum of neutral poly-**8** shows a broad absorbance at λ_{max} = 442 nm which shifts to 766 nm upon oxidation. The spectra of neutral poly-**9** contain broadened bands with λ_{max} at 454. The band maxima for poly-**8** and poly-**9** are red-shifted from those of their corresponding monomers, however they are also blue-shifted with respect to that of poly-**7** (522 nm). This is due to these materials being composed of oligomers with longer conjugation lengths than their corresponding monomers, but with shorter conjugation than in poly-**7**. The observed conductivity in poly-**7** results primarily from charge delocalization along the extended polythiophene chains and π-stacking rather than through the metal bridges in this material.

Our group is interested in conjugated polymers in which coordination of a metal group to an oligothiophene segment may be controllably altered via switching between two different modes, and to examine resulting differences in the structural, chemical, and physical properties of the material. We have prepared Ru(II) bipyridyl phosphinoterthiophene complexes **10** and **11** in which the terminal thiophene is *S*- or *C*-coordinated.[16] Reaction of **10** with NaOH in an acetonitrile-water emulsion resulted in cyclometallation to **11**. Addition of a drop of HPF$_6$ at room temperature to a solution of **11** results in rapid conversion to **10** (Figure 4).

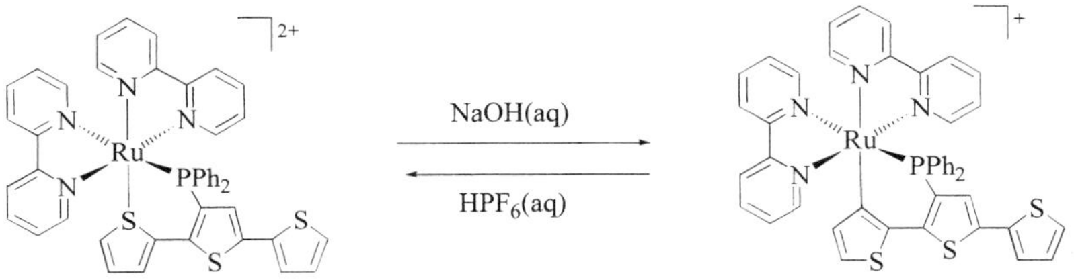

10   **11**

Figure 4. Acid/base switching of thiophene coordination mode in complexes **10** and **11**.

The switch in coordination mode results in significant changes to the $\pi\rightarrow\pi^*$ and MLCT transitions. In the UV-vis absorption spectrum (Figure 2), a shift of the terthiophene $\pi\rightarrow\pi^*$ transition from a shoulder at ~316 nm in **10** to 350 nm in **11** occurs, the Ru d$\rightarrow$bpy π^* MLCT is also red-shifted from 394 nm to 456 nm, and a slight shift was observed for the bpy $\pi\rightarrow\pi^*$ transition.

The Ru$^{2+/3+}$ and terthiophene oxidation potentials of **10** and **11** were measured by cyclic voltammetry. The lowest oxidation waves are assigned to the Ru$^{2+/3+}$ process (quasireversible for **10** when the scan is reversed immediately after the wave). The large shift of the Ru$^{2+/3+}$ oxidation wave of **10** to a substantially lower redox potential in **11** is consistent with the decreased energy of the MLCT transition. The terthiophene oxidation for **10** is most likely the second wave shown, and also shifts to a substantially lower position in **11** (second wave); consistent with the decreased energy of the terthiophene $\pi\rightarrow\pi^*$ transitions.

Although complexes **10** and **11** do not electropolymerize, the synthesis of longer oligomers should allow access to polymeric materials exhibiting the switching observed for these model systems.

Conclusions

A variety of structural motifs have been observed for metal-oligothiophene complexes which can be electropolymerized to give polymeric films. Introduction of ferrocene groups into a conjugated polythiophene backbone results in conductive materials in which charge transfer states arise upon oxidation of the ferrocene groups. Pd-containing polymers in which the Pd is directly bonded to the conjugated backbone were described. Complexes with pendant $Ru(bpy)_2$ groups attached to an oligothiophene backbone via two different coordination modes may be reversible switched using acid and base. Substantial electronic differences between the two coordination modes manifest themselves in changes in the redox and spectroscopic behavior of these complexes. Polymers containing these pendant groups may be useful in sensing applications.

[1] T. A. Skotheim, R. L. Elsenbaumer, J. R. Reynolds, *"Handbook of Conducting Polymers"*, Marcel Dekker, New York 1998.
[2] J. Roncali, *J. Mater. Chem.* **1999**, *9*, 1875.
[3] R. P. Kingsborough, T. M. Swager, *Prog. Inorg. Chem.* **1999**, *48*, 123.
[4] K. Songashira, S. Takahashi, N. Hagihara, *Macromolecules* **1977**, *10*, 879.
[5] M. A. Buretea, T. D. Tilley, *Organometallics* **1997**, *16*, 1507.
[6] I. Manners, *Adv. Organomet. Chem.* **1995**, *37*, 131.
[7] D. L. Compton, T. B. Rauchfuss, *Organometallics* **1994**, *13*, 4367.
[8] M. O. Wolf, *Adv. Mater.,* **2001**, *13*, 545.
[9] G. Tourillon, F. Garnier, *J. Electroanal. Chem.* **1982**, *135*, 173.
[10] J. L. Reddinger, J. R. Reynolds, *Macromolecules* **1997**, *30*, 673.
[11] B. S. Kang, D. H. Kim, T. S. Jung, E. K. Jang, Y. Pak, S. C. Shin, D. S. Park, Y. B. Shim, *Synth. Met.* **1999**, *105*, 9.
[12] Y. Zhu, M. O. Wolf, *Chem. Mater.* **1999**, *11*, 2995.
[13] A. O. Patil, A. J. Heeger, F. Wudl, *Chem. Rev.* **1988**, *88*, 183.
[14] O. Clot, M. O. Wolf, B. O. Patrick, *J. Am. Chem. Soc.* **2000**, *122*, 10456.
[15] O. Clot, M. O. Wolf, B. O. Patrick, *J. Am. Chem. Soc.* **2001**, *123*, 9963.
[16] C. Moorlag, O. Clot, M. O. Wolf, B.O. Patrick, *Chem .Commun.* **2002**, 3028.

Synthesis and Physical Properties of π-Conjugated Metallacycle Polymers of Cobalt and Ruthenium

Masashi Kurashina, Masaki Murata, Hiroshi Nishihara*

Department of Chemistry, School of Science, The University of Tokyo, 7-3-1 Hongo, Bunkyo-ku, Tokyo 113-0033, Japan

Summary: Recent studies on two types of π-conjugated metallacylce polymers are reviewed. Reaction of CpCo(PPh₃)₂ with conjugated diacetylenes afford poly(arylene cobaltacyclopentadienylene) and that of CpRuBr(cod) does poly(arylene ruthenacyclopentrienylene)s in ambient conditions. Regioselectivity of the former metallacycling reacion is not perfect (at most 80% of the 2,5-diaryl selectivity) but that of the latter is satisfactory (~100% of the 2,5-diaryl selectivity) for the formation of π-conjugated structure. Electrochemical oxidation of the cobaltacyclopentadiene polymer and reduction of the ruthenacycle polymer occur facilely and quasi-reversibly by the contribution of metal *d*-orbitals. Physical properties in undoped (neutral) and doped (charged) sates show the behavior of electronic band structure derived from the organic π-conjugated main chain strongly coupled with the metal *d*-orbitals. This affords, for example, photoconductivity in the neutral form of the cobaltacylopentadiene polymer and ferromagnetic interaction in the reduced form of the ruthenacyclopentatriene polymer.

Keywords: cobalt; conducting polymers; metallacycle; organometallic polymers; polymer; redox properties; ruthenium

Introduction

"Organometallic conducting polymers"[1] are derived from the combination of transition metal complexes with π-conjugated organic polymer chains, so-called "conducting polymers", which exhibit various curious physical and chemical characteristics such as facile oxidation and reduction leading to charge storage, high electronic conductivity in the doped states, electrochromism, photo- and electro-luminescence.[2] Various kinds of organometallic conducting polymers and oligomers such as polymetallocenylene,[3,4] poly(metalyne),[5-9] poly(metallo-phthalocyanine),[10,11] polydecker sandwich compounds,[12] thiolate complex polymers,[13-15] cyclobutadienecobalt complex polymers,[16] and others have been synthesized.[17]

Another category of such polymers is poly(arylene metallacyclopentadienylene), wherein the framework of the polymer is π-conjugated and in part composed of metallacyclopentadienes. In this polymer, the metallacyclopentadiene ring with its d-block heteroatom is structurally analogous to the rings found in other representative π-conjugated polymers, such as poly(pyrrole) and poly(thiophene), which contain p-block heteroatoms. Our synthesis of the first metallacyclopentadine polymers, poly(arylene cobaltacyclopentadiene)s, has been accomplished by a new type of polymerization scheme, MetallaCycling Polymerization (MCP), that is based on successive metallacyclization of conjugated diacetylenes.[18] Zirconacyclopentadiene polymers using MCP reactions have been reported by Tilley et al.[19] We have recently succeeded in the preparation of ruthenacyclopentatriene-based organometallic polymers using the similar MCP reactions.[20] This is the first example of the organometallic polymers involving metallacyclopentatriene and thus interesting to compare the properties with metallacyclopentadiene polymers. Our studies on the synthesis and physical properties of cobaltacyclopentadiene polymers have been overviewed in a separate paper,[21] and here we describe mainly the comparison of the synthesis and physical properties between the cobaltacyclopentadiene and ruthenacyclopentatriene polymers.

Synthesis

1. Poly(arylene cobaltacylopentadienylene)s

It is well established that the addition of two acetylenes to the metal center affords metallacyclopentadiene, which is the precursor of the formation of six-membered aromatic rings.[22-25] In the case of cobaltacyclopentadiene, the most common starting compound is $CpCo(PPh_3)_2$ ($Cp = \eta^5\text{-}C_5H_5$), which reacts with two acetylenes, $CR^1{\equiv}CR^2$ and $CR^3{\equiv}CR^4$ stepwise, affording $CpCo(CR^1{\equiv}CR^2)(PPh_3)$ and then $CpCo(CR^1{=}CR^2\text{-}CR^3{=}CR^4)(PPh_3)$ (Equation (1)). A further addition reaction with unsaturated compounds such as acetylene and nitrile gives aromatic six-membered rings. It is also known that the thermal reaction of cobaltacyclopentadiene $CpCo(CR^1{=}CR^2\text{-}CR^3{=}CR^4)(PPh_3)$ affords a cyclobutadiene complex, $CpCo(\eta^4\text{-}C_4R^1R^2R^3R^4)$ (Equation (1)).[26] This kind of unique chemical reactivity of metallacyclopentadiene units can be utilized to obtain interesting polymeric substances from metallayclopentadiene polymers. The regioselectivity of the metallacyclization is important to obtain highly π-conjugated polymers by MCP. When the cobaltacyclopentadiene unit is formed from two CR≡CR' molecules, there are three possible isomers, $2,4-R_2-3,5-R'_2$, $2,5-R_2-3,4-R'_2$ and $3,4-R_2-2,5-R'_2$ forms. Wakatsuki et al. have shown the rule that the acetylenic carbon bearing a bulky group becomes the α-carbon of the metallacyclopentadiene.[27]

Application of the metallacyclization to the polymer synthesis has been successfully made using conjugated diacetylene with the formula, HC≡C-Ar-C≡CH (Ar = 1,4-phenylene, 2-fluoro-1,4-phenylene, 2,5-difluoro-1,4-phenylene, and 4,4'-biphenylene). The diacetylenes HC≡C-Ar-C≡CH of the first attempt have been chosen because the insertion of the arylene moieties in the diacetylene is because the diacetylene without a spacer, RC≡C-C≡CR, could not yield polymers because of the high steric hindrance due to Cp and PPh₃ groups around the cobalt center,[28,29] and because they would give $2,5-Ar_2$ structures according to the stereochemical rule of metallacyclization as noted above. The respective polymers, **1 – 4** were obtained as insoluble powders,[30] and also as films when the glass plates were immersed in the solution of the MCP synthesis. It should be noted that the results on the soluble polymers suggest incomplete π-conjugation by the involvement of $2,4-Ar_2$ units (vide infra), while the extension of π-conjugation by the involvement of $2,5-Ar_2$ structures in the polymers was supported by the electronic spectra of the polymers showing the red shift of the absorption edge compared with the monomeric complex.

Improvement of the solubility was carried out using hexylcyclopentadienyl (HexCp) ligand instead of Cp in the starting cobalt complex[31,32] and/or using alkyl-terminated diacetylenes, $RC\equiv C\text{-Ar-}C\equiv CR$ (R = Me, Bu).[33,34] An MCP reaction between $(HexCp)Co(PPh_3)_2$ and p-diethynylbenzene at 4 °C for 4 days gave no insoluble product, and the GPC spectrum of the reaction product indicated several oligomeric and polymeric components, **5**, the highest molecular weight of which is more than 7×10^4 based on the polystyrene standard.[32] The oligomeric components $HC\equiv C\text{-}C_6H_4[C_4H_2\{Co(HexCp)(PPh_3)\}\text{-}C_6H_4]_nC\equiv CH$ were separated from dimer (n = 2) to nonamer (n = 9). The ^1H NMR analysis indicated that the products were mixtures of these two isomeric structures, 2,5- *vs.* 2,4-substituted cobaltacyclopentadienes with the ratio of 4 : 6. As mentioned above, the regioselectivity is principally dominated by the steric effect of the substituents on the acetylenic carbon. However, this result indicates another factor, probably the dipole-dipole interaction of two acetylenes ligated to cobalt, is concerted in determining the conformation at the intermediate state. Our reinvestigation of the stereochemistry of the metallacyclization for monomeric complexes actually supports this consideration; a reaction of $(HexCp)Co(PPh)_2$ with ethynylbenzene afforded 2,5- and 2,4-diphenylcobaltacyclopentadienes with the ratio of 4 : 6. The yield of 2,5-diarylcobalta-cyclopentadiene was increased by using 1-propynylbenzene instead of ethynylbenzene. This acetylene can generate the large steric repulsion between phenyl and methyl groups in the 2,4-diphenyl-3,5-dimethylcobaltacyclopentadiene, and actually afforded the desirable isomer in the yield of 80%.[32] Based on this regioselectivity result, MCP reactions between $(HexCp)Co(PPh_3)_2$ and p-di-1-propynylbenzene were performed at room temperature and 40 °C, affording a soluble product, **6**, of which GPC spectra shows a drastic enhancement of polymerization by the temperature increase.

5: R = H, A = ⟨benzene⟩   6: R = Me, A = ⟨benzene⟩

The reaction of $CpCo(PPh_3)_2$ with $MeC\equiv C\text{-}p\text{-}C_6H_4\text{-}C_6H_4\text{-}p\text{-}C\equiv CMe$,[33] with $MeOCC\equiv C\text{-}p\text{-}C_6H_4\text{-}C_6H_4\text{-}p\text{-}C\equiv CCOMe$,[33] with $BuC\equiv CC_6H_4C_6H_4C\equiv CBu$,[34] with $MeC\equiv C\text{-}p\text{-}C_6H_4\text{-}C\equiv CMe$,[35] and with $MeC\equiv C\text{-}2,5\text{-}C_4H_2S\text{-}C\equiv CMe$[35] afforded soluble polymers, **7** with $M_w/M_n = 4.0 \times 10^5$ ($M_w/M_n = 4.5$), **8** with 3.0×10^4 ($M_w/M_n = 3.7$), **9** with $M_n = 2.7 \times 10^5$ ($M_w/M_n = 5.2$), **10** with $M_n = 3.8 \times 10^5$ ($M_w/M_n = 4.5$), and **11** with $M_n = 4.4 \times 10^3$ ($M_w/M_n = 1.3$), respectively.

It should be mentioned that another method utilizing a polycondensation of a dihalogenated cobaltacyclopentadiene complex, $Cp(PPh_3)[Co\text{-}C(4\text{-}C_6H_4I)=CBu\text{-}CBu=C(4\text{-}C_6H_4I)]$ (**12**), with $Ni(cod)_2$ (cod = cycloocta-1,5-diene), was applied for the synthesis of perfectly π-conjugated polymer, **13**.[34] When the reaction of **12** was carried out with an excess amount of $Ni(cod)_2$ (2.0 eq) at 50 °C, the molecular weight, M_n, reached to 2.0×10^5 ($M_w/M_n = 2.8$) after 12 h. When the reaction of **12** was carried out with an equimolar of $Ni(cod)_2$ at a room temperature, the oligomers up to a hexamer were obtained. The polymer and oligomer could be purified with a recycling preparative GPC method. Especially, a dimer **13₂** and a trimer **13₃** were isolated and used as samples for investigating physical properties in detail.

As mentioned above, it has been reported that thermal tranformation of cobaltacyclopentadienes into cyclobutadienecobalt complexes occures facilely. [26] This implies that polymers composed of cyclobutadienecobalt complex units can be formed in the MCP reactions. Actually, the reaction of CpCo(PPh$_3$)$_2$ with FcC≡C-p-C$_6$H$_4$-C≡CFc affords a cyclobutadienecobalt complex polymer.[36] The reaction at 50 °C for 22 h, followed by reprecipitation from toluene-hexane afforded an orange powdery solid of [p-C$_6$H$_4${(η^4-C$_4$Fc$_2$)CoCp}]$_n$ (14). GPC analysis of 14 showed that M_n = 5 500 and M_w = 9 600, based on the polystyrene standard. The molecular weight of 5 500 indicates that the degree of polymerization is ca. 9.

14

2. Poly(arylene ruthenacylopentatrienylene)s

It has been reported that ruthenacyclopentatriene is formed by the metallacycling reaction of two RC≡CH molecules with CpRuBr(cod) or (η^5-C$_5$Me$_5$)RuCl(cod).[37,38] The difference in the cyclization reaction compared with the cobalt system described above might be the regioselectivity of this reaction, because it was reported that only 2,5-R$_2$ isomer is formed in the case R = Ph in the literature. [37,38] Our study for the reaction using ethynylferrocene also afforded one isomer, 2,5-bis(ferrocenyl)ruthenacyclopentatriene, **15**.[39]

15

The ruthenacyclopentatriene polymer was synthesized as follows.[20] Reaction of (HexCp)RuBr(cod) with 4,4′-diethynylbiphenyl in dichloromethane at 0 °C yielded a reddish-brown polymeric product. This product was purified by washing with ether to remove residual starting materials, and then the residue was extracted with dichloromethane to remove any insoluble components. Evaporation of the solvent under vacuum resulted in a pure product that was a lustrous black film. The product was characterized by elemental analysis, ^{1}H NMR, ^{13}C NMR, IR, and UV-vis spectra, and electrochemical measurements, which together indicated the structure of poly(biphenylene ruthenacyclopentatrienylene), **16**. Polymer **16** is air-sensitive and its thermal stability is low. Its solution turned into an insoluble gel after standing for a few days at room temperature, even under nitrogen.

16

The composition of geometric isomers for ruthenacyclopentatriene units in **16** was examined by ^{1}H NMR spectra. In the spectra of **16** in CD$_2$Cl$_2$, a signal appeared at δ 7.86, assignable to β-protons of ruthenacyclopentatriene,[40] and no signals due to metallacycle α-protons were detected. Two peaks, one at δ 5.12 and the other at 5.05, both attributed to a cyclopentadienyl ligand, indicate the existence of only one kind of geometric isomers. It is thus concluded that only a 2,5-diaryl derivative of ruthenacyclopentatriene exists in **16**. The result that no geometrical isomer was formed in the polymerization indicates the formation of a fully π-conjugated main chain structure as noted above, which has not been accomplished by the MCP-formed cobaltacylopentadiene polymers.

Although ruthenacyclopentatriene is known as an intermediate that allows cyclotrimerization of acetylenes to yield a benzene derivative[40] as well as cobaltacycloopentadiene, there was no sign of cyclotrimerization during the polymerization in the conditions of this study. In a reaction of $(\eta^5\text{-}C_5Me_5)RuCl(cod)$ with phenylacetylene, the reaction initially affords ruthenacyclopentatriene, which further reacts with excess acetylene to form $(\eta^5\text{-}C_5Me_5)Ru(1,2,4\text{-triphenylbenzene})$.[38] Although ruthenacycles can react with acetylenes, we found that cyclotrimerization did not occur at 0 ˚C in the reaction of (HexCp)RuBr(cod) with 4-ethynylbiphenyl. In the reaction of (HexCp)RuBr(cod) (0.025 M) with 4-ethynylbiphenyl (0.3 M) at 0 °C, which was monitored by ^{1}H NMR spectra, ruthenacycle **17** was formed in 100% yield, and the amount of **17** did not decrease. As a result, **17** did not react with excess acetylenes under these conditions (Figure 1).

The GPC of **16** in dichloromethane showed that the molecular weight was up to 2×10^4 (the number of units, $n = 40$). The high solubility of **16** can be attained by the attachment of a hexyl group on Cp. The average molecular weights of **16**, M_n and M_w were 3400 ($n = 5.9$) and 5 800, respectively. It was revealed that these values correspond primarily to the theoretical ones by time course GPC analysis of the molecular weight distribution under polymerization at 0 °C (*vide infra*). As the polymerization proceeded, M_n and M_w increased at first and became saturated around 90 h, when the M_w/M_n ratio approached the value of 2. In the polycondensation mechanism, the reaction of (HexCp)RuBr(cod) with 4,4′-diethynylbiphenyl is in equilibrium with the reaction of **16** with COD, which is eliminated from (HexCp)RuBr(cod), and both M_n and M_w are maximal when the initial concentrations of (HexCp)RuBr(cod) and 4,4′-diethynylbiphenyl are equal. The maximal M_n and M_w depend on the equilibrium constant of the polycondensation.

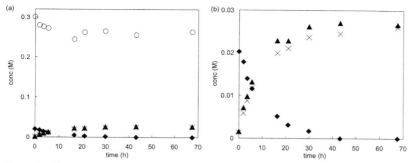

Figure 1. Time course change of concentration in reaction of (HexCp)RuBr(cod) with 4-ethynylbiphenyl (♦: (HexCp)RuBr(cod), ○: 4-ethynylbiphenyl, ▲: **17**, ×: COD). The reaction at 0 °C (a) and its enlarged view (b). (Reprinted with permission from ref. 20)

We assumed that the reaction of (HexCp)RuBr(cod) with 4,4'-diethynylbiphenyl is in equilibrium with the reaction of **16** with COD where k_1 and k_{-1} are the rate constants for forward and back polymerization. If the polymerization was polycondensation and initial concentration of (HexCp)RuBr(cod) and 4,4'-diethynylbiphenyl are the same ($[A]_0$), the number average polymerization degree, $\bar{P}_n$, and the weight average polymerization degree, $\bar{P}_w$, are given by Equation (2) and (3)[41]

$$\bar{P}_n = 1 + \sqrt{K}\,\frac{1-\exp\!\left(-2k_1[A]_0 t/\sqrt{K}\right)}{1+\exp\!\left(-2k_1[A]_0 t/\sqrt{K}\right)} \tag{2}$$

$$\bar{P}_w = 2\bar{P}_n - 1 \tag{3}$$

where K is the equilibrium constant (k_1 / k_{-1}) and t is the reaction time. For $[A]_0 = 0.025$ M, $K = 15$, and $k_1 = 4.4 \times 10^{-4}$ M^{-1}s^{-1}, Eqs (2) and (3) are in agreement with the time course change of M_n and M_w in polymerization. Calculated curves of M_n and M_w are plotted on Figure 2. At equilibrium (t → ∞), the maximum of the $\bar{P}_n$ is given by Equation (4).[37]

$$\overline{P}_n(t \to \infty) = 1 + \sqrt{K} \tag{4}$$

The maximum values are calculated to $M_n = 2\,800$ and $M_w = 4\,800$ from Equation (4). These values are slightly smaller than those experimentally obtained as noted above. This could be because reprecipitation to purify polymer **16** removed the shorter polymer.

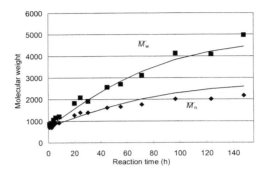

Figure 2. Time course change of M_n and M_w in reaction of (HexCp)RuBr(cod) with 4,4'-diethynylbiphenyl. (Reprinted with permission from ref. 20)

Physical properties

1. Poly(arylene cobaltacylopentadienylene)s

1-1. Electronic spectra

The color of all the cobaltacyclopentadiene polymers noted above is dark brown and the UV-Vis absorption spectra of the films of **1-4** coated on quartz glass show strong bands that have an edge at 500 to 600 nm.[32] As for the soluble polymer, the band edge shifts to the higher wavelength according to an increase in polymerization degree for **9** in CH_2Cl_2 solutions, indicating the formation of π-conjugated structure.[32] The perfectly π-conjugated polymer **13** prepared by polycondensation on dihalogenated complex exhibit a shift of the peak edge to the longer wavelength compared with the corresponding polymer, **9** prepared by metallacycling polymerization (Figure 3).[34]

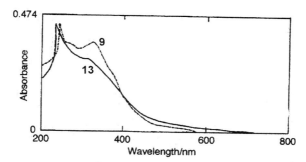

Figure 3. The UV/VIS spectra of compounds **9** and **13** in CH_2Cl_2. (Reprinted with permission from ref. 34)

Band gap energies, E_g, for the cobaltacyclopentadiene polymers were evaluated from the absorption edge based on the semiconductor theory.[36] The E_g values thus evaluated were 2.1 to 2.3 eV,[32] which correspond roughly to the value observed in poly(thiophene) (2.0 eV) and are relatively small when compared to the band gaps among previously known π-conjugated organic polymers.[2]

1-2. Redox properties

Cobaltacylopentadiene complexes undergo one-electron oxidation and their potential and chemical reversibility depend strongly on the substituents on the metallacycle.[42-45] This directly reflects the redox properties of the cobaltacyclopentadiene polymers. The oxidation potential of the polymers becomes more positive when the electron-withdrawing substituents such as –COMe were bound to the metallacycle,[33] and the chemical reversibility increases in the order of the substituents, -COMe < -H < alkyl (-Me, Bu).[33] The polymers with alkyl groups as substituents show cyclic voltammograms indicating high chemical reversibility at a scan rate of 0.1 Vs^{-1} in NBu_4ClO_4-CH_2Cl_2. These results support our consideration that the HOMO based on the *d*-orbitals of the metal atoms in the polymer exists between the valence band (VB) and the conduction band (CB) derived from π-conjugation, and the oxidation occurs at metal sites.

In the cyclic voltammograms of a dimer 13_2 and a trimer 13_3, the waves are broader compared with that of the monomer, although only a single oxidation peak is observed for the dimer and the trimer (Figure 4).[34] This discrepancy suggests that Co(III) and Co(IV) sites are weakly interacted and the mixed-valence states, [Co(III), Co(IV)], [Co(III), Co(IV), Co(III)] and [Co(IV), Co(III), Co(IV)], are generated within a narrow potential range. On the basis of computer simulation, the oxidation potentials are calculated to be $E^{0'}_1 = -0.285$ V and $E^{0'}_2 = -0.212$ V vs. ferrocenium/ferrocene (Fc^+/Fc) for the dimer, and $E^{0'}_1 = -0.291$ V, $E^{0'}_2 = -0.248$ V, and $E^{0'}_3 = -0.189$ V vs. Fc^+/Fc for the trimer. The $\Delta E^{0'}$ values indicate that the intereaction energy between ferrocene and ferrocenium sites, u_{OR} is estimated to be -2 kJ mol^{-1} based on the neighboring site intereaction model of Aoki and Chen,[46] and the value is about one-fifth compared with that for oligo(1,1'-dihexylferrocenylene)s.[47,48]

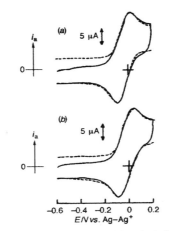

Figure 4. Cyclic voltammograms of compounds 13_2(a) and 13_3(b) at a glassy carbon electrode in 0.1 M NBu_4ClO_4-CH_2Cl_2 at a scan rate of 0.1 Vs^{-1}(full lines) and their simulation based on the open boundary finite-diffusion model(broken lines) . (Reprinted with permission from ref. 34)

The oxidation wave in the cyclic voltammograms of thenylene-bridged cobaltacyclopentadiene polymer, **11**, is fairly broader than that of the phenylene-bridged one.[35]   This is because the energy level for the highest occupied π-orbital of thiophene is closer than that of phenylene to *d*-orbital level of the cobalt site, so that the internuclear electronic interaction through the thiophene ring is considered to be stronger.   In the oxidation process, there are more than one oxidation wave due to the formation of mixed-valence states overlap, resulting in a broad wave in the cyclic voltammogram.

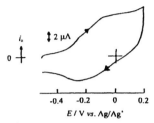

Figure 5. Cyclic voltammogram of a cobaltacyclopentadiene polymer complex, **11** at a glassy carbon electrode in 0.1 M NBu$_4$ClO$_4$-CH$_2$Cl$_2$ at a scan rate of 0.1 Vs^{-1}. (Reprinted with permission from ref. 35)

1-3. Electrical conductivity and photoconductivity

The cobaltacyclopentadiene polymers **4** and **7** show an electrical conductivity of 10^{-12} – 10^{-6} Scm^{-1} in the neutral form at room temperature.[33,49]   When **7** was treated with I$_2$, the conductivity increased up to 10^{-4} Scm^{-1}.[33] This result could be interpreted by the consideration that I$_2$-doping generates Co(III)/Co(IV) mixed-valence states in the polymer chain; this state is stable to some extent as suggested by the electrochemical properties; and consequently, Co(III) and Co(IV) sites are interacted through a π-conjugated chain, causing the mixed-valence conductivity.

The intrinsic photoconductive property was found for a cobaltacyclopentadine polymer **6**.[32] Photo-response of i-V characteristics for ITO/**6**/ITO indicates that the polymer has a low conductivity in dark, and photocurrent is four times larger than the dark current. This kind of remarkable photoconductivity does not appear for common π-conjugated organic polymers in the undoped state but is caused by forming charge-transfer complexes with donor or acceptor

154

molecules such as fullerene.[50,51] We propose that the metal d-character orbitals localized at cobalt sites and their energy level lying between valence and conduction bands act as the trapping sites of holes generated by photo-activation of electrons from the valence band to the conduction band.

2. Poly(arylene ruthenacylopentatrienylene)s

2-1. Electronic spectra

UV-vis-NIR absorption spectra of the monomeric ruthenacyclopentatriene complex **17** and the ruthenacyclopentatriene polymer **16** in dichloromethane are displayed in Figure 6.[20] The bands of **17** at 256, 378, and 505 nm are shifted to 296, 400, and 525 nm, respectively, for **16**. These red shifts should be caused by the extension of the conjugated system for the polymer. However, the band at 694 nm remains unchanged between **16** and **17**. This can be ascribed to LMCT (carbene $\pi \rightarrow$ Ru d) at the ruthenacyclopentatriene unit.

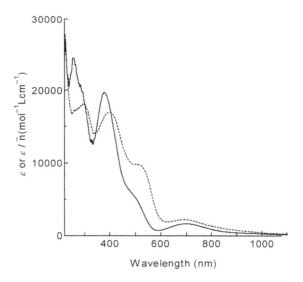

Figure 6. The UV-vis-NIR spectra of **17** (solid line) and **16** (dotted line). (Reprinted with permission from ref.[20])

© 2004 WILEY-VCH Verlag GmbH & KGaA, Weinheim

2-2 Redox properties

Ruthenacyclopentadiene complexes undergo reversible $1e^-$ reduction as shown in the cyclic voltammogram of **17,** indicating the role of ruthenacyclopentatriene as a good electron acceptor (Figure 7). This redox property is quite different from those of cobaltacyclopentadiene, for which the oxidation potential is 0–0.5 V vs. Fc^+/Fc, indicating that it functions as a fine electron donor.[32] The cyclic voltammogram of the polymer **16** in Figure 7 indicates that it also undergoes reversible $1e^-$ reduction ascribed to the ruthenacyclopentatriene unit at $E^{0\prime} = -1.01$ V for **16** and -1.03 V for **17** vs. Fc^+/Fc. The peak of **16** is broader than that of **17,** indicating an existence of electronic interaction between ruthenacycle units in **16**. The location of the modestly localized Ru-centered orbital between the π and π^* orbitals of the conjugated chain is consistent with the electronic spectra as noted above.

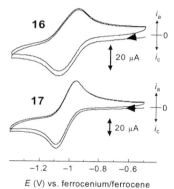

Figure 7. Cyclic voltammograms of **16** (top) and **17** (bottom) on a glassy carbon disk in 0.1 M Bu_4NClO_4-CH_2Cl_2 at a scan rate of 0.1 Vs^{-1}. (Reprinted with permission from ref.[20])

To evaluate the ability of the ruthenacycle to assist the interaction between the ferrocenyl moieties, cyclic voltammetry was carried out with the 2,5-diferrocenyl derivative, **15** in Bu_4NClO_4-dichloromethane, the result of which is shown in Figure 8.[39] A reversible reduction wave of the ruthenacycle was observed at -1.40 V vs. Fc^+/Fc (-1.18 V vs. Ag/Ag^+), and two quasi-reversible waves of the ferrocene moieties appear at -0.01 and 0.23 V vs. Fc^+/Fc. The separation of the oxidation waves by the ferrocenyl moieties in **15** indicates the existence of an interaction between the two ferrocene moieties. The difference in potential of the two waves, 0.24 V, is smaller than those for simple biferrocene (0.42 V) [52] and cobaltacyclopentadienyl-bridged complex **18** (0.47 V),[45] whereas the value is larger than those for most bis-ferrocenyl compounds with π-conjugated organic bridges, such as 1,2-diferrocenylethylene (0.17 V)[53] and 1,2-diferrocenylacetylene (0.13 V).[52]

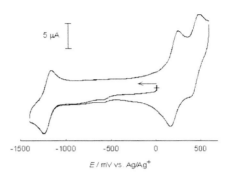

Figure 8. Cyclic voltammogram of **15** on a glassy carbon disk in 0.1 M Bu$_4$NClO$_4$-CH$_2$Cl$_2$ at a scan rate of 0.1 V s^{-1}. (Reprinted with permission from ref.[39])

To investigate the delocalization of electron in a mixed-valence state, **15** was oxidized with 1 eq of [Fe(C$_5$H$_4$Cl)$_2$]PF$_6$ [54] in dichloromethane, and the electronic spectrum was measured (Figure 9). In response to the oxidation, the absorption bands up to 800 nm were blue-shifted, and a new broad band, assignable to the intervalence-transfer between the Fe(II) and Fe(III) atoms, appeared in the near-IR region. Deconvolution of the spectrum into Gaussian functions revealed that the near-IR band was centered at 1180 nm (= 8484 cm^{-1}) with FWVH = 3840 cm^{-1}. Based on Marcus-Hush theory,[55] the mixing coefficient was calculated to be 0.06, which is smaller than or comparable to those for biferrocene (0.09),[52] 1,2-diferrocenylethylene (0.09),[53] and 1,2-diferrocenylacetylene (0.07).[52] From these measurements, the ruthenacyclopentatriene moiety of **15** is found to conduct significant electronic interaction between its 2- and 5-substituents like π-conjugated organic bridges, although the magnitude of this interaction appears to be smaller than that of the cobaltacyclopentadiene analogue **18**. One possible explanation for the larger internuclear interaction in **18** than in **15** is that the stronger donor ability of the cobaltacycle than the ruthenacycle enhances the hole transfer in the electron exchange between the ferrocenyl moieties. However, it may also be due to the difference in the π-conjugation on the carbon atoms.

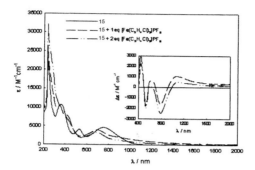

Figure 9. Electronic spectra of **15**. and its oxidized forms generated by addition of 1 eq and 2 eq. [Fe(C$_5$H$_4$Cl)$_2$]PF$_6$ in CH$_2$Cl$_2$. Inset: differences in the spectra of the oxidized forms from the spectrum of **15**. (Reprinted with permission from ref.[39])

18

2-3. Physical Properties in the Reduced State

It is interesting to see the magnetic interaction between Ru sites via π-conjugated linker in the reduced form of **16**. We investigated this interaction by comparing the EPR spectra between the reduced forms of **16** and **17**.[20] Reductions of **16** and **17** were carried out under vacuum by treatment with cobaltocene vapor. To confirm whether or not cobaltocene reduced **16** and **17** successfully, we measured the changes in the UV-vis-NIR spectra of **16** and **17** after adding cobaltocene. We found that the spectrum of **17** was recovered by using iodine vapor to reoxidize the reduced form of **17** (Figure 10). This indicates the reversibility of the chemical redox reaction. For the reduced form of **16**, 20% of spectrum **16** was recovered by oxidation with iodine at room temperature. The changes in the UV-vis-NIR spectra of **16** and **17** showed the same tendency; that is, the band at ca. 500 nm decreased, and the LMCT band at $\lambda_{max} = 694$ nm shifted to ca. 800 nm.

The EPR spectra of the reduced forms of **16** and **17** in frozen THF are displayed in Figure 11. The reduced form of **17** afforded axial symmetry with weak rhombic spectra ($g_\perp = 2.18$, $g_{//} = 1.99$). The spectra was similar to that of an $S = 1/2$ spin in a low-spin d^5 configuration of Ru(III).[56-58] Oxidation number of Ru in the reduced form of **17** is formally V because that in the ruthenacyclopentatriene is VI (carbene counted as dianion). Spin of the Ru in the reduced form of **17** in d^3 configuration is $S = 1/2$ because the strong ligand field and the low symmetry of the complex structure divides the d-orbitals into non-degenerated components. Instability of the reduced form of **17** as well as that of **16** makes difficult the measurement of magnetic susceptibility to confirm S. Hyperfine interaction with ^{99}Ru and ^{101}Ru isotopes ($I = 5/2$, 12.7% and 17%, respectively) was resolved in the spectra, especially for g_2 (a = 3 mT), indicating that the unpaired electron of the reduced form of **17** localizes in the ruthenacycle moiety and does not interact with other molecules. On the other hand, the EPR spectrum of the reduced form of **16** showed several peaks in a wide magnetic field and was completely different from that of **17**. If there is no interaction between Ru sites, the spectrum should be the same as that of **16**, and if there is antiferromagnetic interaction between Ru sites, the signals should diminish. However, neither occurred in this case. We attribute this inconsistency to the ferromagnetic interaction between Ru sites. The interaction was not inter-molecular but intra-molecular, because the concentration of the reduced form of **16** (0.03 mM, converted to ruthenacycle unit concentration) was much lower than that of **17** (0.27 mM). The signal at $g = 2$ was caused mainly by isolated spin in the ruthenacycle, but it should be the sum of this state and the triplet state, indicated by a signal at $g = 4$. Broad signals at 50, 205, 553, and 855 mT indicated interactions between more than two ruthenacyclopentatriene units.

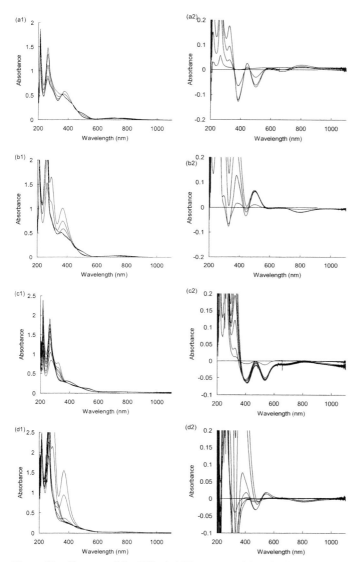

Figure 10. Changes of the UV-vis-NIR spectra with reduction and oxidation of **16** and **17**. Reduction of **17** and their oxidation, and reduction of **16** and their oxidation are displayed in (a1)-(d1) respectively. Differences in spectra of (a1)-(d1) from that of each initial state are displayed in (a2)-(d2) respectively. (Reprinted with permission from ref. 20)

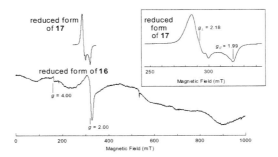

Figure 11. The EPR spectra of reduced forms of **17** (top) and **16** (bottom) at 4 K. Inset: enlarged view of EPR spectrum of reduced form of **17**. (Reprinted with permission from ref.[20])

Conclusion

The metallacyclization reaction using conjugated diacetylenes affords π-conjugated conducting organometallic polymers. Regioselectivity of the metallacyclization is important for the π-conjugation, and up to present, 80% of the regioselectivity has been achieved for the cobaltacylopentadiene polymers and 100% for the ruthenacyclopentatriene polymer. The oxidation of the cobalt center in the cobaltacyclopentadiene polymer occurs facilely and the location of the Co *d*-orbital between the conduction and valence bands brings intrinsic photoconductivity. The reduction of the ruthenium center in the ruthenacyclopentatriene polymer occurs facilely, and the ferromagnetic interaction between the ruthenium centers are observed in the reduced form. These results indicate the dominant role of metal centers in the synthesis and physical properties of π-conjugated metallacycle polymers. In other words, new properties and fuctions are expected by the development of new metallacycle polymers.

Acknowledgments

This work was supported in part by Grants-in-Aid for Scientific Research (Nos. 14044021 (area 412), 14204066) from the Ministry of Culture, Education, Science, Sports, and Technology, Japan, and by The 21st Century COE Program for Frontiers in Fundamental Chemistry.

[1]   H. Nishihara, in *"Handbook of Organic Conductive Molecules and Polymers"*, H. S. Nalwa, Ed., Wiley, Weinheim 1997, Vol. 2, Chapter 19, pp. 799-832.

[2]   H. S. Nalwa (Ed.), *"Handbook of Organic Conductive Molecules and Polymers"*, Wiley-VCH, Weinheim 1997.

[3]   E. W. Neuse, *J. Macromol. Sci.-Chem.* **1981**, *A16*, 3.

[4]   T. Yamamoto, K. Sanechika, A. Yamamoto, *Inorg. Chim. Acta* **1983**, *73*, 75.

[5]   Y. Okamoto, M. C. Wang, *J. Polym. Sci., Polym. Let. Ed.* **1980**, *18*, 249.

[6]   K. Krikor, M. Rotti, P. Nagles, *Synth. Met.* **1987**, *21*, 353.

[7]   H. Matsuda, H. Nakanishi, M. Kato, *J. Polym. Sci., Polym. Lett. Ed.* **1984**, *22*, 107.

[8]   N. Hagihara, K. Sonogashira, S. Takahashi, *Adv. Polym. Sci.* **1980**, *41*, 159.

[9]   M. S. Khan, S. J. Davies, A. K. Kakkar, D. Schwartz, B. Lin, B. F. D. Johnson, J. Lewis, *J. Organomet. Chem.* **1992**, *87*, 424.

[10]   J. W. P. Lin, L. P. Dudek, *J. Polym. Sci., Polym. Chem. Ed.* **1985**, *23*, 1579.

[11]   S. Venkatachalam, K. V. C. Rao, P. T. Manoharan, *Synth. Met.* **1988**, *26*, 237.

[12]   T. Kuhlmann, S. Roth, J. Rozière, W. Siebert, *Angew. Chem. Int. Ed. Engl.* **1986**, *25*, 105.

[13]   R. A. Clark, K. S. Varma, A. E. Underhil, J. Becher, H. Toftlund, *Synth. Met.* **1988**, *25*, 227.

[14]   J. R. Reynolds, F. E. Karasz, C. P. Lillya, J. C. W. Chien, *J. Chem. Soc., Chem. Commun.* **1985**, 268.

[15]   R. Vincete, J. Ribas, P. Cassoux, L. Valade, *Synth. Met.* **1986**, *13*, 265.

[16]   M. Altmann, U. H. F. Bunz, *Angew. Chem. Int. Ed. Engl.* **1995**, *34*, 569.

[17]   H. S. Nalwa, *Appl. Organometal. Chem.* **1990**, *4*, 91.

[18]   A. Ohkubo, K. Aramaki, H. Nishihara, *Chem. Lett.* **1993**, 271.

[19]   J. R. Nitschke, S. Zurcher S, T. D. Tilley, *J. Am. Chem. Soc.*, **2000**, *122*, 10345.

[20]   M. Kurashina, M. Murata, T. Watanabe, and H. Nishihara, *J. Am. Chem. Soc.*, **2003**, *125*, 12420.

[21]   H. Nishihara, M. Kurashina, M. Murata, *Macromolecular Symposia*, **2003**, *196*, 27.

[22]   H. Yamazaki, N. Hagihara, *Bull. Chem. Soc. Jpn.* **1971**, *44*, 2260.

[23]   Y. Wakatsuki, H.Yamazaki, *J. Chem. Soc., Chem. Commun.* **1973**, 280.

[24]   Y. Wakatsuki, T. Kuramitsu, H. Yamazaki, *Tetrahedron Lett.* **1974**, 4549.

[25]   H. Bönnenman, *Angew. Chem., Int. Ed. Engl.* **1985**, *24*, 248.

[26]   K. M. Nicholas, M. O. Nestle, D. Seyferth, in *"Transition Metal Organometallics in Organic Synthess"*, H. Alper Ed., Academic Press, New York, 1978, Vol. 2.

[27]   Y. Wakatsuki, O. Nomura, K. Kitaura, K. Morokuma, H. Yamazaki, *J. Am. Chem. Soc.* **1983**, *105*, 1907.

[28]   T. Shimura, A. Ohkubo, K. Aramaki, H. Uekusa, T. Fujita, S. Ohba, H. Nishihara, *Inorg. Chim. Acta* **1995**, *230*, 215.

[29]   T. Fujita, H. Uekusa, A. Ohkubo, T. Shimura, K. Aramaki, H. Nishihara, S. Ohba, *Acta Cryst.* **1995**, *C51*, 2265.

[30]   H. Nishihara, T. Shimura, A. Ohkubo, N. Matsuda, K. Aramaki, *Adv. Mater.* **1993**, *5*, 752.

[31]   N. Matsuda, T. Shimura, K. Aramaki, H. Nishihara, *Synth. Metals* **1995**, *69*, 559.

[32]   T. Shimura, A. Ohkubo, N. Matsuda, I. Matsuoka, K. Aramaki, H. Nishihara, *Chem. Mater.* **1996**, *8*, 1307.

[33]   I. Matsuoka, K. Aramaki, H. Nishihara, *Mol. Cryst. Liq. Cryst.*, **1996**, *285*, 199.

[34]   I. Matsuoka, K. Aramaki, H. Nishihara, *J. Chem. Soc., Dalton Trans.*, **1998**, 147.

[35]   I. Matsuoka, H. Yoshikawa, M. Kurihara, H. Nishihara, *Synth. Metals* 1999, *102*, 1519.

[36]   M. Murata, T. Hoshi, I. Matsuoka, T. Nankawa, M. Kurihara, H. Nishihara, *J. Inorg. Organomet. Polym.* **2000**, *10*, 209.

[37]   M. O. Albers, D. J. A. deWaal, D. C. Lies, D. J. Robinson, E. Singleton, M. B. Wiege, *J. Chem. Soc., Chem. Commun.* **1986**, 1681.

[38]   C. Ernst, O. Walter, E. Dinjus, S. Arzberger, H. Görls, *J. Pract. Chem.* **1999**, *8*, 341.

[39]   Y. Yamada, J. Mizutani, M. Kurihara, H. Nishihara, *J. Organomet. Chem.* **2001**, *80-83*, 637.

[40]   T. Naota, H. Takaya, S. Murahashi, *Chem. Rev.* **1998**, *98*, 2599.

[41]   S. Nakahama, T. Nose, S. Akiyama, K. Sanui, Y. Tsujita, M. Doi, K. Horie, K. *Essential Koubunshi Kagaku*; Koudan-Sha: Tokyo, 2000; Chapter 3, 27.

[42]   R. S. Kelly, W. E. Geiger, *Organometallics* **1987**, *6*, 1432.

[43]   B. T. Donovan, W. E. Geiger, *J. Am. Chem. Soc.* **1988**, *110*, 2335.

162

[44] B. T. Donovan, W. E. Geiger, *Organometallics* **1990**, *9*, 865.
[45] A. Ohkubo, T. Fujita, S. Ohba, K. Aramaki, H. Nishihara, *J. Chem. Soc., Chem. Commun.* **1992**, 1553.
[46] K. Aoki, J. Chen, *J. Electroanal. Chem.*, **1995**, *380*, 35.
[47] T. Hirao, M. Kurashina, K. Aramaki, H. Nishihara, *J. Chem. Soc., Dalton Trans.*, **1996**, 2929.
[48] H. Nishihara, T. Hirao, K. Aramaki, K. Aoki, *Synth. Metals*, **1997**, *84*, 935.
[49] H. Nishihara, A. Ohkubo, K. Aramaki, *Synth. Metals*, **1993**, *55*, 821.
[50] K. Yoshino, S. Morita, T. Kawai, H. Araki, X. H. Yin, A. A. Zakhidov, *Synth. Met.* **1993**, *56*, 2991.
[51] N. S. Sariciftci, L. Smilowitz, D. Braun, G. Srdaniov, V. Srdanov, F. Wudl, A. J. Heeger, *Synth. Met.* **1993**, *56*, 3125.
[52] C. Lavenda, K. Bechgaard, D. O. Cowan, *J. Org. Chem.,* **1970**, *41*, 2700.
[53] A. C. Ribou, J. P. Launay, M. L. Sachtleben, H. Li, C. W. Spangler, *Inorg. Chem.,* **1996**, *35*, 3735.
[54]   T. Hirao, K. Aramaki, H. Nishihara, *Bull. Chem. Soc. Jpn.,* **1998**, *71*, 1817.
[55]   N. S. Hush, *Coord. Chem. Rev.,* **1985**, *64*, 135.
[56] A. Pramanik, N. Bag, G. K. Lahiri, A. Chakravorty, *J. Chem. Soc., Dalton Trans.* **1990**, 3823.
[57] A. Ceccanti, P. Diversi, G. Ingrosso, F. Laschi, A. Lucherini, S. Magagna, P. Zanello, *J. Organomet. Chem.* **1996**, *526*, 251.
[58] D. Bhattacharyya, S. Chkraborty, P. Munshi, G. K. Lahiri, *Polyhedron* **1999**, *18*, 2951.

Macromol. Symp. **2004,** *209,* 163-176

Synthesis and Lithographic Applications of Highly Metallized Cluster-Based Polyferrocenylsilanes

Wing Yan Chan, Alison Y. Cheng, Scott B. Clendenning, Ian Manners*

Department of Chemistry, University of Toronto, 80 St. George Street, Toronto, Ontario, Canada, M5S 3H6
E-mail: imanners@chem.utoronto.ca

Summary: We report the use of a cobalt-clusterized polyferrocenylsilane (Co-PFS) as a precursor to patterned ferromagnetic ceramics. Co-PFS was synthesized. Functioning as a negative resist, Co-PFS lines with widths of 10–300 μm were patterned using UV-photolithography, while features as small as 500 nm were afforded by electron-beam lithography. Subsequent pyrolytic treatment of the lithographically patterned Co-PFS yielded ferromagnetic ceramics containing Fe/Co nanoparticles. Due to its high metal-loading, Co-PFS is a good etch resist for oxygen and hydrogen plasma reactive ion etching. Reactive ion etching of a thin film of Co-PFS in a secondary magnetic field allowed direct access to ferromagnetic ceramic films, providing a viable alternative to pyrolysis.

Keywords: ferromagnetic nanoparticles; lithography; metal clusters; polyferrocenyl-silanes; ring-opening polymerization

Introduction

Patterning of surfaces on the nanometer scale with metals offers the possibility of fabricating materials with useful catalytic, optical, sensing, electrical and magnetic properties. Available patterning methods range from soft lithography, scanning probe lithography, electron-beam (e-beam) lithography (EBL) and photolithography. Recent reports include the use of microcontact printing to order monodisperse nanoparticles of iron oxide,[1] and nanotransfer printing to transfer a gold pattern with 75 nm feature sizes from a gold-coated GaAs stamp to an appropriately primed polydimethylsiloxane substrate.[2] Alternatively, scanning probe techniques also offer precise control of patterning. An atomic force microscope (AFM) tip has been used to pen 35 nm-wide lines of MoO_3 through local oxidation of a Mo film,[3] while 30 nm-wide Pt lines have been drawn via the reduction of H_2PtCl_6 at an AFM tip using electrochemical AFM dip-pen

© 2004 WILEY-VCH Verlag GmbH & KGaA, Weinheim · · · · · DOI: 10.1002/masy.200450511

lithography.[4] In EBL, a film of an organometallic polymer can be used as a resist. Johnson and co-workers have used thin films of the organometallic cluster polymer [Ru$_6$C(CO)$_{15}$Ph$_2$PC$_2$PPh$_2$]$_n$ as a negative resist in EBL to direct write conducting wires (ca. 100 nm wide) composed of metal nanoparticles.[5] This example illustrates the convenience and utility of combining conventional lithographic techniques with easily processible organometallic polymer resists. Moreover, post-development treatments of the patterned organometallic resist such as pyrolysis or reactive ion etching (RIE) offer additional control over the chemical and physical properties of the surface.

While polymers containing metal clusters are attractive as lithographic resists, their synthesis remains a poorly developed area. Nevertheless, three examples are worth mentioning. The previously mentioned organometallic polymer with ruthenium carbonyl clusters in the backbone [Ru$_6$C(CO)$_{15}$Ph$_2$PC$_2$PPh$_2$]$_n$,[5] was prepared by reacting [Ru$_6$(CO)$_{17}$] with [Ph$_2$PC$_2$PPh$_2$] in tetrahydrofuran under reflux to give the dark brown polymer. Its degree of polymerization was estimated to be 1 000–1 020 by electron microscopy. In the second example, Humphrey et al. used a step-growth polycondensation approach to synthesize oligourethanes containing bimetallic Mo$_2$Ir$_2$ clusters in the backbone (Figure 1).[6] In the presence of catalytic amounts of dibutyltin diacetate (DBTA), the bimetallic cluster diol [Mo$_2$Ir$_2$(CO)$_{10}${η-C$_5$H$_4$(CH$_2$)$_2$OH}$_2$] was reacted with 1,ω-alkyl- or aryldiisocyanates to give red powders or waxy solids. The degree of polymerization was estimated using gel permeation chromatography and ^{1}H NMR spectroscopy; this ranged from 5–30, depending on the length of the spacer. While the diisocyanate alkyl group chain length had little effect on the extent of polymerization, increasing the spacer length between the cyclopentadienyl and hydroxyl groups led to a significant increase in M_n.

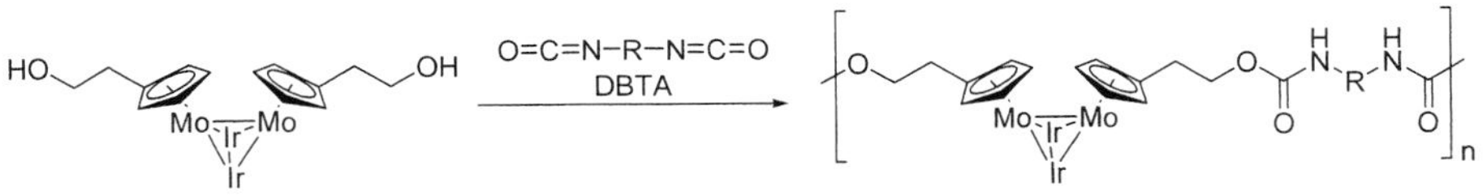

Figure 1. Synthesis of bimetallic cluster polymers (Mo$_2$Ir$_2$ = Mo$_2$Ir$_2$(CO)$_{10}$, R = alkyl or aryl spacer). (Adapted from [6])

In the third approach, Brook *et al.* synthesized oligomers and polymers containing alkynylsilane and arylsilane groups; they were complexed with either dicobalt octacarbonyl at the alkyne, or chromium hexacarbonyl at the aryl group to give metallized oligomers or polymers.[7] Using the same routes, a highly metallized oligomer containing both chromium and cobalt was prepared via sequential clusterization of both aryl- and alkynylsilane groups (Figure 2).

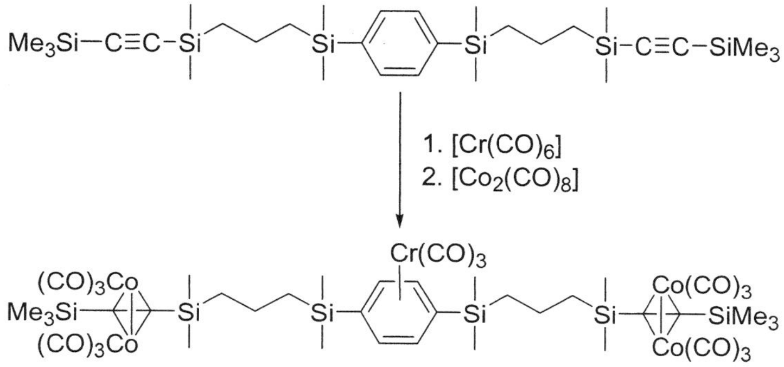

Figure 2. Structure of an oligosilane containing chromium and cobalt carbonyl clusters. (Adapted from [7])

One focus of our current research is the incorporation of metal clusters into a well-defined organometallic polymer that can be lithographically patterned. Thermal,[8] anionic[9] and transition metal-catalyzed[10] ring-opening polymerization (ROP) of sila[1]ferrocenophanes (**1**) are well-established routes to high molecular weight, soluble polyferrocenylsilanes (**2**, PFS) that contain covalently bonded iron atoms in the main chain (Scheme 1). The incorporation of PFS into patterned surfaces has already yielded materials with tunable magnetic properties that may find applications as protective coatings, magnetic recording media and anti-static shields.[11] Furthermore, polymers containing organometallic moieties exhibit low plasma etch rates compared to their purely organic counterparts,[12] suggesting they can be used as etch masks to deposit interesting materials.

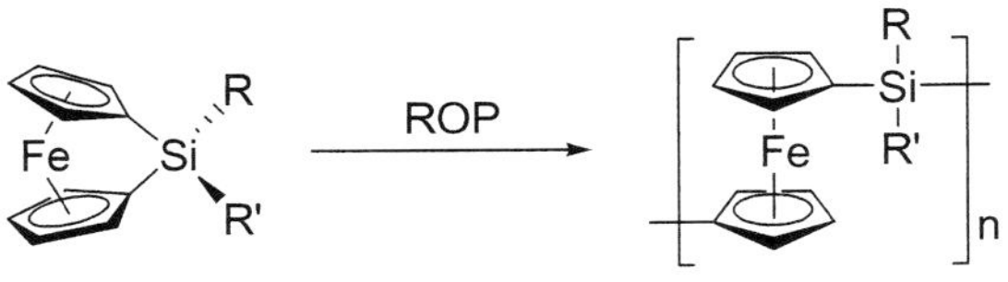

Scheme 1. ROP of a sila[1]ferrocenophane (1) to afford a polyferrocenylsilane (2).

Results and Discussion

Synthesis of Highly Metallized PFS and Precursors to Metal-Containing Ceramics

The introduction of additional metals into the PFS chain can increase metal loadings and allow access to binary or higher metallic alloy species. This can be accomplished by adding an acetylenic substituent at silicon that can be further transformed via reaction with an appropriate metal carbonyl. The synthesis of an acetylide-substituted sila[1]ferrocenophane and the corresponding highly metallized PFS is shown in Scheme 2. 1,1'-Dilithioferrocene was reacted with trichloromethylsilane to form the chloro-substituted sila[1]ferrocenophane 3. Lithium phenylacetylide was then used to substitute the chloride group of 3 to yield 4. Monomer 4 was ring-opened with a platinum(0) catalyst to give high molecular weight polymer 5.[13] The acetylenic substituent in 5 was clusterized with dicobalt octacarbonyl to yield the highly metallized PFS (Co-PFS) (6) with three metal atoms per repeat unit.[14]

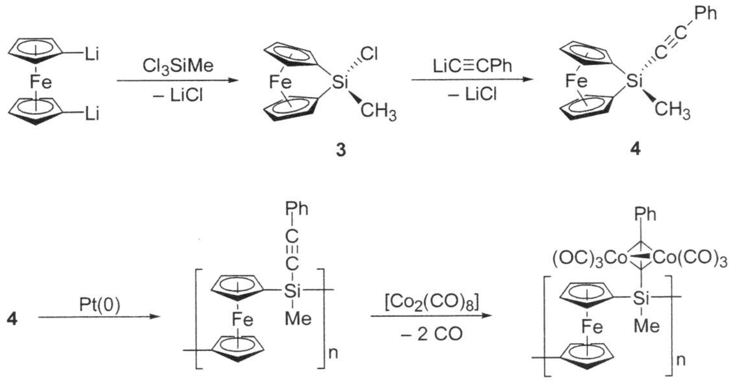

Scheme 2. Synthesis of highly metallized Co-PFS (6).

We tested the utility of Co-PFS as a magnetic ceramic precursor.[14] Pyrolysis of **6** under a nitrogen atmosphere at either 600 °C or 900 °C afforded black magnetic ceramics in relatively high yields (72% and 59%, respectively). Powder X-ray diffraction studies revealed that the ceramic residue contained Fe/Co alloy particles embedded in an amorphous C/SiC matrix. Figure 3 shows the cross-sectional transmission electron microscopy (TEM) images of the ceramics prepared at 600 °C and 900 °C. Electron-rich metal nanoparticles can clearly be seen, and their chemical compositions were determined by electron energy-loss spectroscopic (EELS) elemental mapping experiments. The results indicated that both iron and cobalt were localized in the same nanoparticles, thereby suggesting a homogeneous alloy. Superconducting quantum interference device (SQUID) magnetometry showed that ceramics formed at 900 °C were ferromagnetic with no blocking temperature up to 355 K.

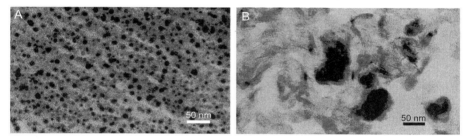

Figure 3. Cross-sectional TEM images of ceramics resulting from pyrolysis of Co-PFS **6** at (a) 600 °C and (b) 900 °C

Highly Metallized Sila[1]ferrocenophanes as Polymer Precursors

Although highly metallized PFS can be obtained via functionalization of the polymer, this reaction is often incomplete as the core of the polymer may be well shielded from reagents. To circumvent this problem, the monomer could be functionalized and then ring-opened to give a fully functionalized polymer. We have investigated the reactions between acetylide-substituted monomer **4** and metal complexes containing cobalt, molybdenum and nickel (Scheme 3).[15]

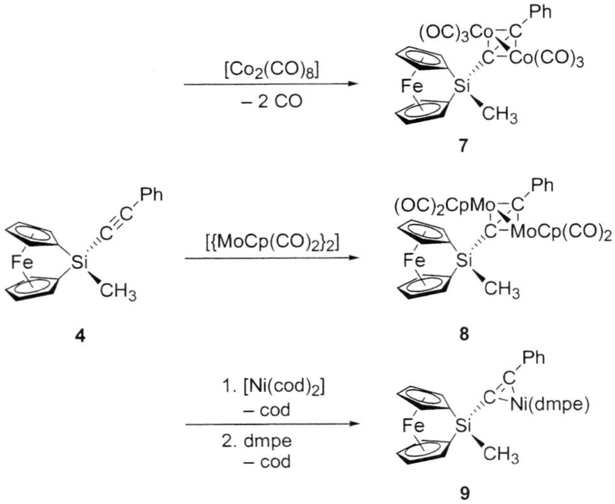

Scheme 3. Synthesis of sila[1]ferrocenophanes containing pendent cobalt, molybdenum, or nickel units.

As in the synthesis of Co-PFS **6**, the triple bond in monomer **4** reacted smoothly with dicobalt octacarbonyl to give the cobalt-clusterized product **7**. Molybdenum clusters were introduced via molybdenum cyclopentadienyl dicarbonyl dimer; this reaction was sluggish at room temperature and required heating to 75 °C to form **8**. Bis(1,5-cyclooctadiene)nickel(0) also reacted with the triple bond in **4** to add a nickel unit; however, all attempts to separate the desired product from unidentified side-products were unsuccessful, possibly due to subsequent dissociation of the labile cod ligand. When 1,2-bis(dimethylphosphino)ethane (dmpe) was added to the above product, substitution of the cod ligand occurred to give nickel-containing sila[1]ferrocenophane **9**. Investigations into the ROP behaviour of these highly-metallized sila[1]ferrocenophanes are underway.

Having demonstrated that different metal complexes could be incorporated into the acetylenic substituent of **4**, we turned our attention to the synthesis of a platinum-containing sila[1]ferrocenophane. After ring-opening polymerization, the Pt-PFS formed may be used as a magnetic ceramic precursor to FePt nanocrystals. FePt nanoparticles are currently under investigation for ultrahigh-density magnetic recording media applications due to their large uniaxial magnetocrystalline anisotropy.[16] We attempted the reaction of acetylide-substituted monomer **4** with tris(triethylphosphine)platinum(0).[15] Instead of reacting with the triple bond, [Pt(PEt₃)₃] inserted into the *ipso*-cyclopentadienyl carbon–silicon bond to form platinasila[2]ferrocenophane **10** (Scheme 4). The tendency of platinum(0) fragments to undergo oxidative insertion has so far prevented the synthesis of a sila[1]ferrocenophane with a pendent platinum unit.

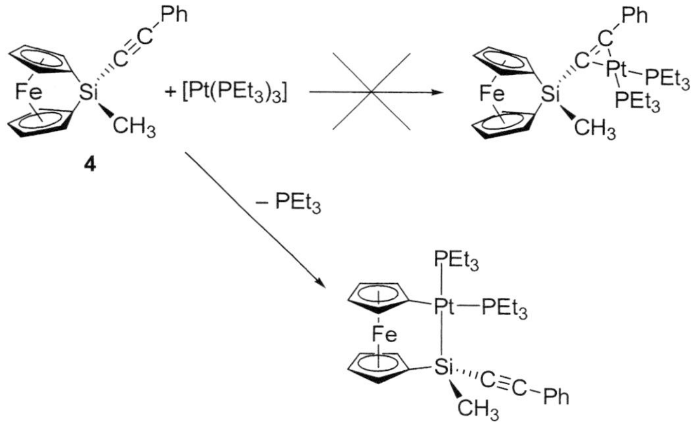

Scheme 4. Attempted synthesis of a platinum-containing sila[1]ferrocenophane; formation of platinasila[2]ferrocenophane **10**.

Lithography

The use of Co-PFS **6** as a lithographic resist offers several advantages: high metal content, ease of processibility, atomic level mixing and stoichiometric control over composition. We have recently demonstrated the use of this metallopolymer as a resist in electron-beam lithography (EBL),[17] UV-photolithography,[18] as well as O₂- and H₂-RIE (RIE = reactive ion etching).[19]

Cluster-Based Polyferrocenylsilanes as EBL Resists

In order to determine whether Co-PFS **6** could function as an e-beam resist, uniform thin films (*ca.* 200 nm thick) of **6** were spin-coated onto silicon substrates and EBL was carried out at various doses and currents in a modified scanning electron microscope. The treated films were then developed in THF and characterized. Co-PFS was found to be a negative e-beam resist, with best results obtained at a dose of 25 mC/cm². Shapes including dots and bars were successfully fabricated (Figure 4).

Figure 4. SEM images of (a) dots and (b) bars fabricated by EBL using a Co-PFS **6** resist

The elemental composition of these patterns was investigated using time-of-flight secondary ion mass spectrometry (TOF SIMS) and X-ray photoelectron spectroscopy (XPS). Elemental maps of iron and cobalt for the array of bars on the silicon substrate were obtained using TOF SIMS. The mapping clearly revealed that iron and cobalt were concentrated within the bars. Information regarding the chemical environment and distribution of iron and cobalt throughout the bars was obtained via XPS and compared to data from an untreated film of Co-PFS. The atomic ratio of Fe:Co for the bars was in agreement with the theoretical value of 1:2. Detailed scans for iron and cobalt revealed no change in binding energy for the elements at 3 nm and 12 nm depths, indicating uniformity in the average chemical environment. Overall, XPS revealed little change in the chemical composition of the polymer resist after EBL. Finally, magnetic force microscopy (MFM) indicated no appreciable magnetic field gradient above the bars following EBL.

In order to enhance the magnetism of the patterned bars, the were pyrolyzed at 900 °C under a nitrogen atmosphere to promote the formation of metallic nanoclusters.[14,20] The same array of bars was characterized before and after pyrolysis by tapping mode AFM. Comparison of the images and cross-sectional profiles suggested there is excellent shape retention, accompanied by a decrease in the dimensions of the bars. MFM studies of the bars indicated that they contained heterogeneous ferromagnetic clusters whose magnetic dipoles appeared to be randomly oriented. The ferromagnetism of the pyrolyzed array of ceramic bars was independently confirmed by magneto-optic Kerr effect (MOKE) measurements.

Although the aforementioned proof-of-concept experiments dealt with the formation of micron-scale objects,[17] EBL can be used to routinely fabricate structures down to 30–50 nm.[21] Indeed, this process has already been extended to the patterning of sub-500 nm bars and dots from Co-PFS (**6**).

Cluster-Based Polyferrocenylsilanes as UV Photoresists

Organic polymers incorporating acetylenic moieties have been shown to crosslink under thermal conditions. Upon heating, the acetylene groups from adjacent chains undergo cyclotrimerization and coupling reactions, creating crosslinks in the polymer.[22] In addition, photo-induced polymerization of alkyl- and aryl-substituted acetylenes are known to be catalyzed by metal carbonyls such as $Cr(CO)_6$, $Mo(CO)_6$ and $W(CO)_6$.[23–25] Recently, Bardarau *et al.* demonstrated that polyacrylates with pendent acetylenic side groups could be photocrosslinked with $W(CO)_6$ as catalyst.[26] We believe that Co-PFS **6** is a promising candidate as a resist for UV-photolithography, as it contains both the acetylenic unit and the metal carbonyl catalyst.[14]

To study this possibility, a thin film (*ca.* 200 nm) of Co-PFS **6** on a silicon substrate was exposed to near-UV radiation (λ=350–400 nm, 450 W) for 5 minutes. The exposed film was developed in tetrahydrofuran before characterization. Co-PFS was found to be a negative-tone photoresist; the exposed area is presumably crosslinked and less soluble. This would be consistent with the photo-initiated crosslinking mechanism of acetylenes in the presence of metal carbonyls. However, it is also possible to form crosslinks in Co-PFS through decarbonylation of the cobalt clusters. The thickness of the film before and after UV treatment was determined by ellipsometry. A 200 nm-thick film of Co-PFS had a thickness of *ca.* 170 nm after exposure to UV radiation and solvent development. The decrease in thickness is probably a reflection of the decreased volume of the crosslinked polymer.

Patterning of Co-PFS **6** films was accomplished using a metal foil shadow mask with *ca.* 50-

300 μm features fabricated by micromachining. Figure 5a is an optical micrograph of a straight line of Co-PFS patterned using the shadow mask. Smaller features (*ca.* 10-20 μm) were obtained using a chrome contact mask (Figure 5b). In both cases the unexposed polymer was completely removed during development with THF, leaving behind patterns with well-defined edges.

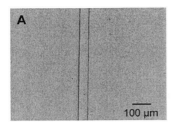 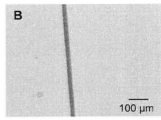

Figure 5. Optical micrographs of Co-PFS **6** lines fabricated by UV photolithography using (a) a shadow mask and (b) a chrome contact mask

A patterned Co-PFS **6** film was pyrolyzed at 900 °C under a nitrogen atmosphere in an attempt to fabricate magnetic ceramic lines. The resulting ceramic lines have the same dimensions as the polymer precursor and show excellent shape retention in the lateral directions. Inspection of the ceramic line at higher magnifications revealed the formation of what appeared to be Co/Fe nanoparticles throughout the line.

UV photolithography using Co-PFS **6** as a resist provides a convenient route to deposit patterned polymer and magnetic ceramic onto flat substrates over large areas. Due to the excellent RIE resistance of the polymer, Co-PFS patterned by UV photolithography can potentially be used for pattern transfer onto the underlying substrate using conventional plasma etching techniques. Investigations are underway to improve the resolution of this resist.

Cluster-Based Polyferrocenylsilanes as Reactive Ion Etch Resists

It has been demonstrated that polyferrocenylsilanes exhibit low plasma etch rates, which can be attributed to the formation of a protective layer of involatile iron and silicon compounds.[27] This property has been exploited with PFS block copolymers that self-assembled into micelles or phase-separated thin films for the deposition of patterned ceramics and pattern transfer onto the substrate. [28–30] In our research, we are interested in the direct formation of high metal content magnetic ceramic films by plasma RIE treatment of Co-PFS.[19]

To test the feasibility of this approach, thin films of Co-PFS **6** on a silicon substrate were treated with either a hydrogen or an oxygen plasma. Effects of the plasmas on the chemical composition of the treated films were studied by TOF SIMS depth profiling. In both cases, plots of the intensity of the Si^+, Fe^+, FeO^+, Co^+ and CoO^+ signals as a function of depth showed that the plasma affected only the top 10 nm of the films, leaving the underlying polymer relatively untouched. The chemical composition of the modified films was analyzed by XPS. In both cases, a Co:Fe ratio of approximately 1:1 was found at 3 nm depth, rather than the stoichiometric 2:1 ratio found in the untreated polymer. This deficiency of Co on the film surface is most likely due to the volatilization of cobalt carbonyl clusters during the high vacuum processing required for RIE. Formation of iron oxides (FeO and Fe_2O_3) as well as cobalt oxides (CoO and Co_3O_4) were observed at 3 nm depth in the oxygen plasma treated films. In the case of H_2-RIE, small amounts of oxides were also observed at 3 nm depth, presumably due to the oxidation of reduced metal upon exposure to atmospheric oxygen.

Morphological changes in the surface of Co-PFS films following RIE were investigated by TEM and AFM. Thin films of the polymer (*ca.* 50 nm thick) on a carbon-coated copper TEM grid were exposed to either an oxygen or a hydrogen plasma. In both cases, analysis by TEM revealed the presence of electron rich nanoworms with widths ranging from 4–12 nm on the supporting carbon film. Untreated samples were featureless, supporting plasma-induced nanoworm formation. Electron energy-loss spectroscopic (EELS) elemental mapping and energy-dispersive X-ray (EDX) analysis indicated that there was a high concentration of iron, cobalt and silicon in the nanoworms. In contrast to the essentially flat and featureless surface of the untreated film, AFM images of both plasma treated films exhibited a pervasive interconnected reticulated structure, which would appear as nanoworms in a projection normal to the plane of the sample. These features are reminiscent of surface reticulations observed by Thomas *et al.* in AFM images of organosilicon polymers following ambient temperature O_2-RIE.[31] We hypothesize that these features resulted from dewetting caused by a polarity difference between the overlying inorganic layer and the native polymer, or spinodal decomposition of the strained inorganic layer.

174

In order to access films with useful magnetic properties, plasma-induced crystallization of metallic nanoclusters was attempted on a thin film of Co-PFS **6** in a secondary magnetic field. The Co-PFS films (*ca.* 200 nm thick) on silicon substrates were placed between two samarium-cobalt (SmCo) magnets, aligned with opposite magnetic poles facing each other during H_2- or O_2-RIE. The magnets caused the formation of an intense plasma plume around the sample, which in turn resulted in intense etching conditions. Nanostructures obtained under these conditions were shown to be ferromagnetic by MFM. A tapping mode AFM image and the corresponding MFM image of a hydrogen plasma treated film are shown in Figure 6. These reticulations were much larger than those found for samples treated under similar plasma conditions without the SmCo magnets. We postulate that the secondary magnetic field concentrated the plasma and accelerated nanoworm formation through more efficient removal of carbonaceous material and silicon, while the additional thermal energy present in the plasma plume promoted metal crystallization. To the best of our knowledge, this is the first example of the formation of a ferromagnetic film directly from the plasma treatment of a metallopolymer. A standard lithographic technique such as soft lithography, EBL or UV-photolithography used in conjunction with RIE in a secondary magnetic field should offer access to ordered arrays of ferromagnetic ceramics. These arrays may find applications in spintronics as an isolating, magnetic layer in a nano-granular in-gap structure[32] or the formation of logic circuits using magnetic quantum cellular automata.[33]

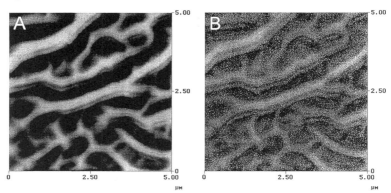

Figure 6. (a) Tapping mode AFM image of a thin film of Co-PFS **6** following H_2-RIE in a secondary magnetic field and (b) the corresponding lift mode MFM image (5×5 μm^2)

Conclusions

We have developed synthetic routes to highly metallized sila[1]ferrocenophanes and the corresponding high molecular weight polymer Co-PFS **6**. This polymer is readily accessible via transition metal-catalyzed ROP of sila[1]ferrocenophane **4**, followed by clusterization of the acetylenic substituents with dicobalt octacarbonyl. The intrinsic high metal content, air-stability and solution processibility of Co-PFS make it an excellent precursor to lithographically patterned ceramics. Fabrication of patterned arrays of polymer and magnetic ceramics on the sub-micron scale was achieved by EBL and pyrolysis of thin films.[17] We also demonstrated that Co-PFS can function as a negative-tone UV-photoresist.[18] The crosslinking mechanism in this system is still under investigation. Finally, we showed the utility of this highly metallized polymer as a RIE resist; treatment of the Co-PFS films with either a hydrogen or an oxygen plasma in the presence of a secondary magnetic field afforded ferromagnetic ceramics.[19]

Acknowledgements

I.M. would like to thank the Canadian Government for a Canadian Research Chair. W.Y.C. acknowledges an NSERC postgraduate scholarship. A.Y.C. thanks the University of Toronto for a Graduate Fellowship in Chemistry. S.B.C. is grateful to NSERC for a PDF. The authors would also like to acknowledge the excellent work of our collaborators: EBL: Prof. Harry E. Ruda; Dr. Stephane Aouba (Center of Advanced Nanotechnology, University of Toronto); AFM/ MFM: Prof. Christopher M. Yip, Mandeep S. Rayat, Guocheng Yang (Department of Chemical Engineering and Applied Chemistry, University of Toronto); TEM: Dr. Neil Coombs, Ellipsometry: Chantal Paquet (Department of Chemistry, University of Toronto); RIE, UV Photolithography, XPS: Prof. Zheng-Hong Lu, Dr. Sijin Han, Dr. Dan Grozea (Department of Materials Science and Engineering, University of Toronto); TOF SIMS: Dr. Rana N. S. Sodhi, Dr. Peter M. Brodersen, (Surface Interface Ontario, Department of Chemical Engineering and Applied Chemistry, University of Toronto); MOKE: Prof. Mark R. Freeman, Jason B. Sorge (Department of Physics, University of Alberta).

[1]   Q. J. Guo, X. W. Teng, S. Rahman, H. Yang, *J. Am. Chem. Soc.* **2003**, *125*, 630.

[2]   Y. L. Loo, R. L. Willet, K. W. Baldwin, J. A. Rogers, *J. Am. Chem. Soc.* **2002**, *124*, 7654.

[3]   M. Rolandi, C. F. Quate, H. J. Dai, *Adv. Mater.* **2002**, *14*, 191.

[4]   Y. Li, B. W. Maynor, J. Liu, *J. Am. Chem. Soc.* **2001**, *123*, 2105.

[5]   B. F. G. Johnson, K. M. Sanderson, D. S. Shephard, D. Ozkaya, W. Z. Zhou, H. Ahmed, M. D. R. Thomas, L. Gladden, M. Mantle *Chem. Commun.* **2000**, 1317.

[6]   N. T. Lucas, M. G. Humphrey, A. D. Rae, *Macromolecules* **2001**, *34*, 6188.

[7]   T. Kuhnen, M. Stradiotto, R. Ruffolo, E. Ulbrich, M. J. McGlinchey, M. A. Brook *Organometallics* **1997**, *16*, 5048.

[8]   D. A. Foucher, B. Z. Tang, I. Manners, *J. Am. Chem. Soc.* **1992**, *114*, 6246.

[9]   Y. Z. Ni, R. Rulkens, I. Manners, *J. Am. Chem. Soc.* **1996**, *118*, 4102.

[10] P. Gómez-Elipe, R. Resendes, P. M. Macdonald, I. Manners, *J. Am. Chem. Soc.* **1998**, *120*, 8348.

[11] [11a] For a review on polyferrocenylsilane chemistry, see: K. Kulbaba, I. Manners, *Macromol. Rapid Commun.* **2001**, *22*, 711; [11b] M. J. MacLachlan, M. Ginzburg, N. Coombs, T. W. Coyle, N. P. Raju, J. E. Greedan, G. A. Ozin, I. Manners, *Science* **2000**, *287*, 1460; [11c] I. Manners *Science* **2001**, *294*, 1664; [11d] K. Temple, K. Kulbaba, K. N. Power-Billard, I. Manners, K. A. Leach, T. Xu, T. P. Russell, C. J. Hawker, *Adv. Mater.* **2003**, *15*, 297.

[12] G. N. Taylor, T. M. Wolf, L. E. Stillwagon, *Solid State Technol.* **1984**, *27*, 145.

[13] A. Berenbaum, A. J. Lough, I. Manners, *Organometallics* **2002**, *21*, 4451.

[14] A. Berenbaum, M. Ginzburg-Margau, N. Coombs, A. J. Lough, A. Safa-Safat, J. E. Greedan, G. A. Ozin, I. Manners, *Adv. Mater.* **2003**, *15*, 51.

[15] W. Y. Chan, A. Berenbaum, S. B. Clendenning, A. J. Lough, I. Manners; *Organometallics* **2003**, *22*, 3796.

[16] S. Sun, C. B. Murray, D. Weller, L. Folks, A. Moser, *Science* **2000**, *287*, 1989.

[17] S. B. Clendenning, S. Aouba, M. S. Rayat, D. Grozea, J. B. Sorge, P. M. Brodersen, R. N. S. Sodhi, Z.-H. Lu, C. M. Yip, M. R. Freeman, H. E. Ruda, I. Manners, *Adv. Mater.* **2004**, in press.

[18] A. Y. Cheng, S. B. Clendenning, S. Han, P. Yang, Z.-H. Lu, C. M. Yip, I. Manners, *unpublished results* **2003**.

[19] S. B. Clendenning, S. Han, N. Coombs, C. Paquet, M. S. Rayat, D. Grozea, P. M. Broderson, R. N. S. Sodhi, C. M. Yip, Z.-H. Lu, I. Manners, *Adv. Mater.* **2004**, in press.

[20] M. Ginzburg, M. J. MacLachlan, S. M. Yang, N. Coombs, T. W. Coyle, N. P. Raju, J. E. Greedan, R. H. Herber, G. A. Ozin, I. Manners, *J. Am. Chem. Soc.* **2002**, *124*, 2625.

[21] J. R. Sheat, B. W. Smith, in: *"Microlithography"*, Marcel Dekker Inc., New York 1998.

[22] S. A. Swanson, W. W. Fleming, D. C. Hofer, *Macromolecules* **1992**, *25*, 582.

[23] T. Masuda, K. Yamamoto, T. Higashimura, *Polymer* **1982**, *23*, 1663.

[24] T. Masuda, Y. Kuwane, K. Yamamoto, T. Higashimura, *Polym. Bull.* **1980**, *2*, 823.

[25] S. J. Landon, P. M. Shulman, G. L. Geoffroy, *J. Am. Chem. Soc.* **1985**, *107*, 6739.

[26] C. Bardarau, Z. Y. Wang, *Macromolecules* **2003**, *36*, 6959.

[27] I. Manners, *Pure. Appl. Chem.* **1999**, *71*, 1471.

[28] J. A. Massey, M. A. Winnik, I. Manners, V. Z. H. Chan, J. M. Ostermann, R. Enchlmaier, J. P. Spatz, M. Möller, *J. Am. Chem. Soc.* **2001**, *123*, 3147.

[29] L. Cao, J. A. Massey, M. A. Winnik, I. Manners, S. Riethmuller, F. Banhart, J. P. Spatz, M. Möller, *Adv. Funct. Mater.* **2003**, *13*, 271.

[30] R. G. H. Lammertink, M. A. Hempenius, J. E. van der Enk, V. Z. H. Chan, E. L. Thomas, G. J. Vancso, *Adv. Mater.* **2000**, *12*, 98.

[31] V. Z. H. Chan, E. L. Thomas, J. Frommer, D. Sampson, R. Campbell, D. Miller, C. Hawker, V. Lee, R. D. Miller, *Chem. Mater.* **1998**, *10*, 3895.

[32] M. B. A. Jalil, *IEEE Tran. Magn.* **2002**, *38*, 2613.

[33] R. P. Cowburn, M. E. Welland, *Science* **2000**, *287*, 1466.

Macromol. Symp. **2004**, *209*, 177-183

Bisglycine-Substituted Ferrocene Conjugates

Heinz-Bernhard Kraatz

Department of Chemistry, University of Saskatchewan, 110 Science Place, Saskatoon, Saskatchewan, Canada S7N 5C9
E-mail: kraatz@skyway.usask.ca

Summary: The solid state structures of three bissubstituted glycine ferrocene conjugates are described allowing a direct comparison of the structural parameters. Whereas the fully protected glycine ester $Fc(Gly-OEt)_2$ adopts a 1,3'-conformation leading exclusively to intermolecular H-bond formation, the free acid $Fc(Gly-OH)_2$ adopts the more compact 1,3'-comformation with intramolecular H-bonding. The same intramolecular H-bonding pattern is adopted by the glycine ferrocenophane $Fc(Gly-CSA)_2$.

Keywords: ferrocene; hydrogen bonding peptide; supramolecular assembly

Introduction

Due to their inherent ability to engage in intermolecular interactions, such as hydrogen bonding or salt-bridge formation, amino acids and peptides are ideally suited as building blocks in the synthesis of supramolecular systems.[1] Recent efforts have been directed at equipping non-covalent supramolecular peptide assemblies with redox-active groups, such as ferrocenes[2] or cobaltocenes and give them specific electric properties that may be exploited for biosensing or may have potential for the design of bioelectronic materials.[3] In particular, 1,1'-disubstituted ferrocene peptide conjugates have proven useful to generate chiral helical supramolecular arrangements that do not rely on intermolecular H-bonding but are the result of chiral patterning. In most cases, this type of ferrocene-peptide conjugates display a rigid intramolecular cross-strand H-bonding interaction in solution and the solid state having the amino group proximal to the Fc group on one strand engages in H-bonding with the CO group of the same amino acid on the opposite peptide strand.[2] Intermolecular H-bonding is only observed if the second amino acid has available H-bonding donor sites, giving an indication of the interplay between inter- and intramolecular H-bonding. Futhermore, the intramolecular H-bonding is robust withstanding even the presence of other strongly H-bonding components that co-crystallize, showing the potential

DOI: 10.1002/masy.200450512

utility of these Fc-conjugates for the systematic supramolecular design of redox active assemblies. We recently reported the use of cystamine-linked Fc-peptide conjugates for the generation of Fc-peptide-modified surfaces. Furthermore, we explored the use of a cystine-linker to connect the podand peptide chains on the two Cp rings in 1,1'-bissubstituted ferrocenes, giving ferrocenophanes. In essence this should restrict the mobility even further providing more rigid building blocks. Recently, the synthesis of peptide ferrocenophanes and their metal chelating abilities was reported.[4] However, structural data are unavailable in the literature. Ferrocenophanes display a range of interesting properties, ranging from metallo-cryptants for the coordination of ions to starting materials for metallopolymers.[5] Our efforts have been guided by our desire to investigate the electron transfer properties in these systems, which should make two peptide-thiol linkages per Fc-peptide conjugate, on the gold surface and compare them to our growing library of simple Fc-peptides.[6] Here we compare the structural features of three bissubstituted ferrocene glycine conjugates: the bisglycine ethylester **1**, the free acid **2**, and the glycine ferrocenophane **3**.

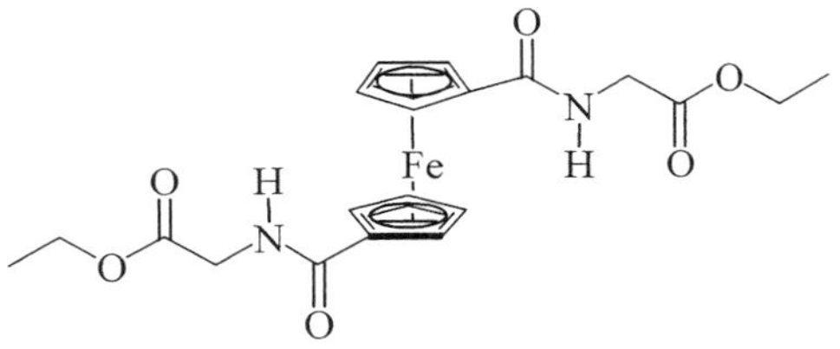

1

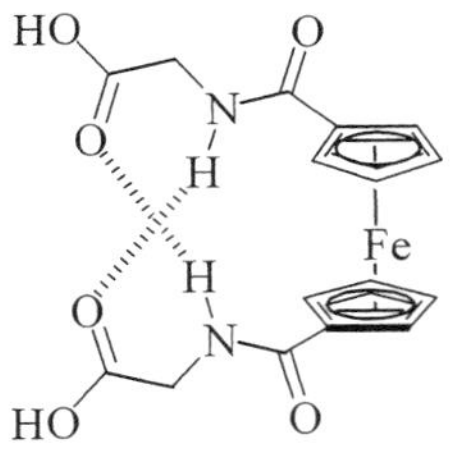

2

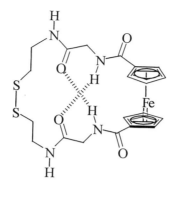

3

Molecular Structure

Single crystals of compounds **1**, **2**, and **3** were obtained by diffusion methods giving yellow to orange x-ray quality crystals. The structure of the glycine ethylester **1** is shown in Figure 1. The compound adopts the for this class of bisubstituted ferrocene conjugates unusual 1,3'-conformation, which allows it to engage in intermolecular H-bonding to neighbouring molecules via the amide O(1) and N(1) with a H-bond distance of 2.839(5) Å. The other podand amino acidester is not involved in any H-bonding. The result is a one dimensional H-bonded zig-zag chain shown in Figure 1 in which the molecules stack alternating with the non-H-bonding glycine ligand pointing up and down. These chains are packed into a layered structure, with an interlayer separation of 4 Å. Similar layer and double layer structures were observed before for Fc-Gly$_2$-Oet, which also form a 1-D H-bonded chain. This H-bonding pattern is reminiscent of that reported by Hirao and coworkers for the monosubstituted Fc-Ala-Pro-OEt with alternating up-down orientation of the molecules.[2a]

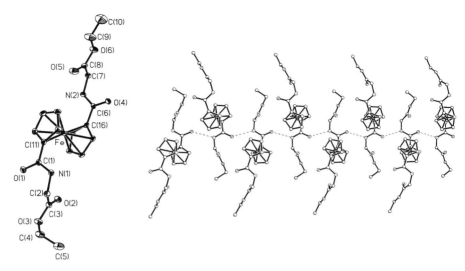

Figure 1. left: Structure of the ferrocenoyl bisglycine ethylester **1** showing the 1,3'-substitution of the ferrocene group. right: Interaction between neighbouring molecules via the interaction of the O(1) of one of the two amide groups with the N(1) of the same amide group of an adjacent molecule resulting in the formation of a 1-D H-bonded polymeric chain.

This creates some degree of asymmetry in the Fc-amide groups, allowing a direct comparison of the effect of H-bonding on the podand glycine ethyester. Althought the difference between the C=O distances of the two amides is small, the one involved in H-bonding is slightly elongated. The difference is clearer in the amide N-C=O distances, in which the H-bonded N(1)-C(1) is significantly shortened compared to the non-H-bonded N-C=O distance (d(N(1)-C(1)) =1.326(5) Å, d(C(6)-N(2)) = 1.344(5) Å). In sharp contrast to the structure of the ester, the free acid, the synthesis of which was reported earlier,[7] shows the now familiar 1,2'-conformations allowing the formation of the cross-strand intramolecular H-bonding interaction involving the two amide NH and the two acid C=O on opposite Cp rings (Figure 2). This interaction also provides the correct conformation for the acid OH and the Fc-C=O to allow additional intermolecular H-bonding giving a two-dimensional H-bonded network. In this network the Fc-C=O of one molecule interacts with the OH groups of an adjacent molecule (d(O(1)···O(3*)) = 2.620(3) Å). Additional weak In O···H-C interactions between O(1) and H(12) are present in the solid state, which contribute to the stability of the network.

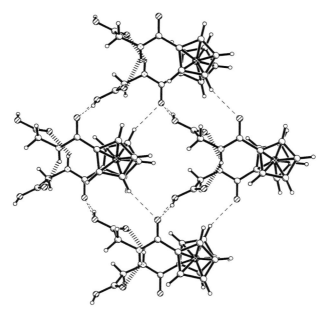

Figure 2. 2-D network formed by the interaction of adjacent molecules of **2**. H-bonding interactions between adjacent molecules involving the interaction of the Fc-C=O group with the acid OH (d(O(1)···O(3*)) = 2.620(3) Å. Individual molecules display the familiar intramolecular H-bonding pattern with d(N-O(2A)) and d(N(A)-O(2)) = 2.875(3) Å.

Compound **3** displays the same basic structural features. The system adopts a 1,2'-configuration similar to the open chain Fc-conjugate **2**, having both Fc-C=O pointing outwards and are engaged in intermolecular H-bonding interactions. Both amide NH in the ferrocenophane **3** are engaged in strong intramolecular H-bonding to the opposite peptide C=O across the ring (see Figure 3), which has been observed in other open-chain ferrocene dipeptide conjugates reported by Hirao and Metzler-Nolte.[2] Importantly, the dihedral angles of the podand peptide chains are slightly influenced by these variations. In particular, the dihedral angle Φ_1 increases somewhat upon de-esterification and ring closure to form the macrocycle.

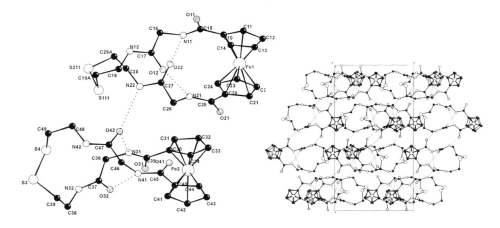

Figure 3. Molecular structure of compounds **3** showing the intra- and intermolecular H-bonding interactions present in the solid state: d(N(11)···O(22)) = 2.862(4)Å, d(N(21)···O(12)) = 2.873(4) Å, d(N(31)···O(42)) = 2.959(4)Å, d(N(41)···O(32)) = 2.815(4) Å; d(N(22)···O(42)) = 2.974(5)Å.

Table 1: Dihedral angles of the podand peptide chain for compounds **1**, **2**, and **3**

	Angle	1	2	3
	Φ_1	65.4(4)°	79.0(2)°	-86.2(4)
		-69.9(5)°		-123.3(5)
				93.7(4)
				115.6(4)
	Ψ_1	-158.4(3)°	-178.60(18)°	176.2(3)
		162.3(3)°		-148.8(4)
				-171.4(3)
				135.9(4)

Conclusion

Deprotection of the glycine ethylester **1** to give the free acid **2** results in a dramatic conformational change that is accompanied by a change in the H-bonding pattern. Whereas **1** forms a 1-D H-bonded polymeric chain, compound **2** forms a polymeric H-bonded 2-D network. Upon closure of the ring to give the ferrocenophane the intramolecular H-bonding pattern remains intact speaking to the robustness of this structural motif. Detailed investigations are currently underway to investigate the use of these ferrocene conjugates as building blocks for the rational assembly of large supramolecular structures.

Acknowledgements

This work was supported by NSERC and the CRC programme.

[1]   [1a] A. Aggeli, I.A. Nyrkova, M. Bell, R. Harding, L. Carrick, T. C. B. McLeish, A. N. Semenov, N. Boden, *Proc. Natl. Acad. Sci. USA* **2001**, *98*, 11857; [1b] W. A. Petka, J. L. Harden, K. P. McGrath, D. Wirtz, D. A. Tirrell, *Science* **1998**, *281*, 389.
[2]   [2a] A. Nomoto, T. Moriuchi, S. Yamazaki, A. Ogawa, T. Hirao, *Chem. Commun.* **1998**, 1963; [2b] T. Moriuchi, A. Nomoto, K. Yoshida, A. Ogawa, T. Hirao, *J. Am. Chem. Soc.* **2001**, *123*, 68; [2c] T. Moriuchi, A. Nomoto, K. Yoshida, T. Hirao, *J. Organomet. Chem.* **1999**, *589*, 50; [2d] T. Moriuchi, K. Yoshida, T. Hirao, *Organometallics* **2001**, *20*, 3101; [2e] T. Moriuchi, K. Yoshida, T. Hirao, *J. Organomet. Chem.* **2003**, *668*, 31; [2f] T. Moriuchi, A. Nomoto, K. Yoshida, T. Hirao, *Organometallics* **2001**, *20*, 1008; [2g] D. R. van Staveren, T. Weyhermüller, N. Metler-Nolte, *Dalton Trans.* **2003**, 210.
[3]   for example: A. Salomon, D. Cahen, S. Lindsey, J. Tomfohr, V. B. Engelkes, C. D. Friesbie, *Adv. Mater.* **2003**, *22*, 1881.
[4]   H. Huang, L. Mu, J. He, J.-P. Cheng, *J. Org. Chem.* **2003**, *68*, 7605.
[5]   see for example: [5a] W. Finckh, B.-Z. Tang, A. Lough, I. Manners, *Organometallics* **1992**, *11*, 2904; [5b] R. Rulkens, D. P. Gates, D. Balaishis, J. K. Puelski, D. F. McIntosh, A. J. Lough, I. Manners, *J. Am. Chem. Soc.* **1997**, *119*, 10976; [5c] K. H. H. Fabian, H.-J. Linder, N. Nimmerfroh, K. Hafner, *Angew. Chem. Int. Ed.,* **2001**, *40*, 3402; [5d] R. Steudel, K. Hassenberg, J. Pickardt, E. Grigiotti, R. Zanello, *Organometallics* **2002**, *21*, 2604; [5e] A. Tarrage, P. Molina, J. L. Lopez, M. D. Velasco, D. Bautista, P. C. Jones, *Organometallics* **2002**, *21*, 2055.
[6]   I. Bediako-Amoa, T.C. Sutherland, C.-Z. Li, R. Silerova, H.-B. Kraatz, *J. Phys. Chem. B* **2004**, *108*, 704.
[7]   G. Schachschneider, M. J. Wenzel, *J. Labelled Comp.* **1985**, *12*, 235.

Synthesis of Iron-Containing Polymers with Azo Dyes in the Backbones or Side Chains

Alaa S. Abd-El-Aziz,* Rawda M. Okasha, Erin K. Todd, Tarek H. Afifi, Patrick O. Shipman, K. Michelle D. Copping

Department of Chemistry, The University of Winnipeg, Winnipeg, Manitoba, Canada R3B 2E9
E-mail: a.abdelaziz@uwinnipeg.ca

Summary: The synthesis of cationic cyclopentadienyliron-containing polymers with pendent azobenzene chromophores was accomplished via metal-mediated nucleophilic aromatic substitution reactions. All of the desired polymers were isolated as vibrantly coloured materials and displayed excellent solubility in polar aprotic solvents. Cationic and neutral cyclopentadienyliron polymers incorporating azo dyes in the backbone were also prepared. Reactions of azo dyes with dichlorobenzene complexes allowed for the isolation of cationic cyclopentdienyliron (CpFe$^+$) complexes with azo dye chromophores. These complexes were then reacted with 1,1'-ferrocenedicarbonyl chloride to produce the trimetallic monomers with terminal chloro groups. These monomers contained two pendent CpFe$^+$ cations and a neutral iron moiety in the backbone. Nucleophilic substitution reactions of these monomers with oxygen and sulfur containing dinucleophiles gave rise to a new class of polymeric materials. The pendent CpFe$^+$ moieties could also be cleaved from the polymer backbones using photolysis to afford novel ferrocene based polymers. The UV-vis spectra of the organoiron monomers and polymers display similar wavelength maxima however incorporating azobenzene chromophores with electron-withdrawing substituent into the polymer chains resulted in bathochromic shifts of the λ_{max} values.

Keywords: arene complexes; azo dyes; ferrocene-based polymers; iron-containing polymers; nucleophilic substitution reactions; polyethers; polythioethers

© 2004 WILEY-VCH Verlag GmbH & KGaA, Weinheim   DOI: 10.1002/masy.200450513

Introduction

Interest in the field of organometallic polymers stems from the wide spread applications that these materials find, and the synthesis of organoiron polymers is a major contributor in these studies.[1-7] The first reports of ferrocene-based polymers in 1955 initiated a number of research groups to develop various routes to prepare organometallic polymers.[8] The past twenty years have seen a significant increase in the synthesis of these polymers in light of their potential applications as electrocatalysts, modified electrodes, chemical sensors and photoactive molecular devices.[5-7, 9-11] Our group has focused on the use of cyclopentadienyliron coordinated chloroarenes in the synthesis of cationic iron-based polymers.[12-16] The mild reaction conditions and the use of various nucleophiles has allowed for the synthesis of many functionalized monomeric and polymeric ethers, thioethers, amines and imines.[12-15] It has been documented that the incorporation of azo dyes into the backbones or the side chains of organic and organometallic macromolecules generates interesting properties such as liquid crystallinity and non-linear-optical activity. The properties that this class of polymers possess make them potential candidates for a wide variety of applications such as reversible optical storage systems, electrooptic modulators, and photorefractive switches.[17-27] Manners and coworkers have prepared liquid crystalline poly(ferrocenyl silanes) containing azo dyes in their side chains.[28] These polymers were prepared via thermal ring opening polymerization of a methyl silyl-bridged [1]ferrocenophane which was subsequently functionalized with azobenzene moieties through a platinum-catalyzed hydrosilylation reaction.[28] This article will focus on the synthesis of cationic iron-coordinated polyaromatic ethers and thioethers functionalized with azo dyes in their side chains as well as the incorporation of azo dyes directly into the polymeric backbones.

Results and Discussion

Polymers with Pendent Azobenzene Moieties

Recently, our group has been focusing on the preparation of organometallic and organic polyethers, thioethers and amines using metal-mediated nucleophilic aromatic substitution reactions. Condensation reactions of bimetallic organoiron complexes containing terminal carboxylic acid groups with a phenolic substituent on an azo dye in the presence of dicyclohexylcarbodiimide (DCC) and N,N-dimethylamino pyridine (DMAP), allowed for the isolation of brightly coloured azobenzene substituted diiron complexes. These complexes were then polymerized via nucleophilic aromatic substitution reactions with S- and O-based nucleophiles to yield polyaromatic ethers and ether/thioethers with pendent azobenzene chromophores **1a-c** to **3a-c** (Scheme 1). The functionalized polymers were isolated in good yields as bright orange-red solids. These materials displayed good solubility in polar organic solvents such as acetonitrile, DMF and DMSO. The electron-withdrawing properties of the R groups on the azobenzene moieties gave rise to bathochromic shifts of the λ_{max} values. For example, the λ_{max} of polymers **1a-c** were 417, 452 and 489 nm, respectively.

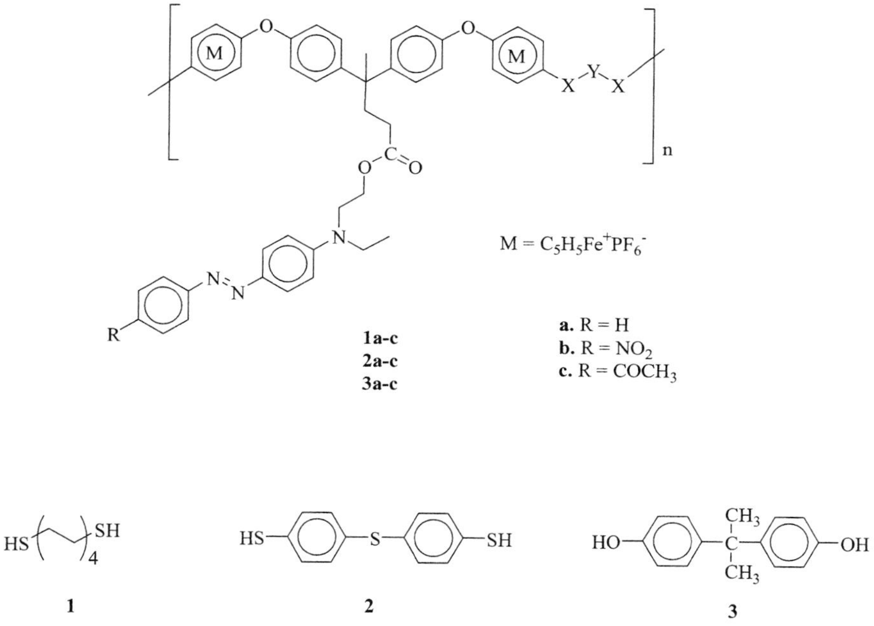

$M = C_5H_5Fe^+PF_6^-$

1a-c
2a-c
3a-c

a. R = H
b. R = NO_2
c. R = $COCH_3$

1   **2**   **3**

Scheme 1

Characterizations of the monomeric and polymeric materials were conducted using spectroscopic and analytical techniques. Figure 1 shows the ^1H and ^{13}C NMR spectra of polymer **2a**. In the ^1H NMR spectrum, the uncomplexed aromatic protons appear as four doublets and two sets of multiplets and resonate between 6.85 and 7.73 ppm, while the complexed aromatic protons appear as a singlet at 6.26 ppm. The cyclopentadienyl protons show as an intense singlet at 5.14 ppm and the remaining methyl and methylene protons appear between 4.22 and 1.12 ppm. The ^{13}C-NMR spectrum was run as an attached proton test (APT). The two methyl carbons peaks resonate at 12.32 and 27.30 ppm and the methylene carbons resonate between 30.22 and 61.95 ppm. The cyclopentadienyl carbons appear as a large peak at 79.00 ppm, and the complexed aromatic carbons can be seen at 76.52 and 85.43 ppm, however, the hydrogen-attached aromatic carbons resonate between 111.79 and 135.16 ppm.

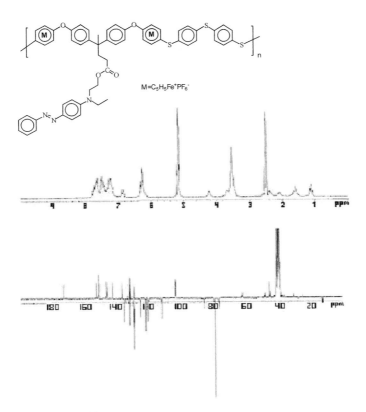

Figure 1. ^{1}H-NMR and ^{13}C-NMR spectra of organoiron polymer **2a**.

The molecular weights of the polymers were determined using gel permeation chromatography (GPC). Photolytic demetallation of the organoiron polymers was performed prior to molecular weight deremination in order to avoid interactions of the metal moieties with the GPC columns. The weight average molecular weights of the metallated polymers were estimated to be in the range of 13,400 to 18,200 with polydispersities between 1.2 to 2.0.

Thermal properties of both the metallated and demetallated polymers were analyzed using thermal gravimetric analysis (TGA) and differential scanning calorimetry (DSC). The thermograms of the metallated and demetallated polymers containing azobenzene groups in their side chains showed multiple weight losses. TGA analysis of the metallated polymers showed that the weight losses between 220-240 °C correspond to the cleavage of the cyclopentadienyliron moieties. The decomposition of the azo group was in the range of 250-290 °C, however the degradation of the polymer backbones was in the range of 420-460 °C. The DSC analysis showed that the glass transition temperatures (T_g's) of the metallated polymers occurred between 167 and 173 °C, whereas the demetallated polymers exhibited glass transition temperatures between 111 and 123 °C. It was observed that the T_g's of the metallated polymers were higher than the T_g's of the demetallated ones. This observation was attributed to the presence of the bulky cationic cyclopentadienyliron moieties. Furthermore, the organoiron polymers with aromatic spacers possessed higher T_g's than the polymers containing aliphatic spacers; this trend was also observed in the demetallated polymers.

Azobenzene Moieties in the Polymer Backbone

Synthesis of the monomers occurred through nucleophillic aromatic substitution of the *p*-dichlorobenzene cyclopentadienyliron complex with azo dyes in the presence of K_2CO_3 and DMF. These vibrantly coloured azobenzene complexes were isolated in good yields, and the azobenzene complexes were subsequently reacted with 1,1' ferrocenedicarbonyl chloride in dichloromethane (Scheme 2) to form triiron monomers with terminal cationic cyclopentadienyliron moieties and an inner disubstituted bis(cyclopentadienyliron) group (**6a-c**).

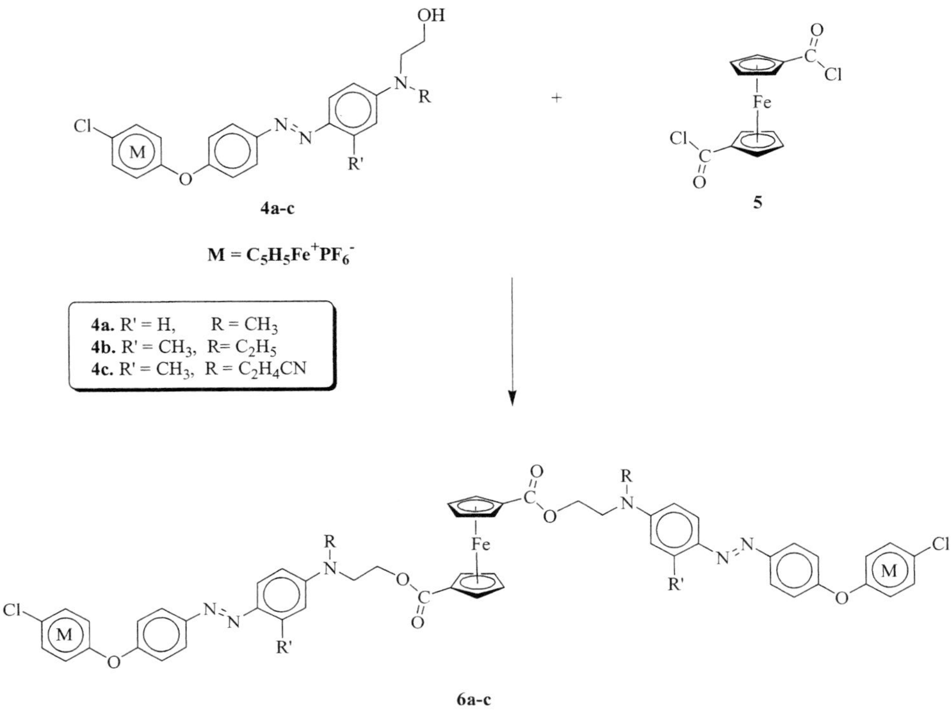

Scheme 2

Characterization of all complexes was preformed using ^{1}H-NMR, ^{13}C-NMR, and IR spectroscopy. For example, the ^{1}H-NMR spectrum of monomer **6a** shows that the ferrocenyl protons appear at 4.4 and 4.8 ppm and the cyclopentadienyl protons of the cationic complex resonate as a singlet at 5.41 ppm. The complexed aromatic protons shift upfield between 6.6 and 6.9 ppm, while the uncomplexed aromatic protons appear between 7.5 and 8.0 ppm. Polymerization of the azobenzene functionalized complexes, accomplished with oxygen and sulfur based nucleophiles, allowed for the isolation of novel organoiron polymers containing azobenzene chromophores in the backbones (Scheme 2). The polymerization was carried out at 50 °C for 7 h, and the resulting brightly coloured polymers displayed good solubility in polar organic solvents.

Cleavage of the cationic iron moieties was performed via photolytic demetallation reactions that resulted in the isolation of the ferrocene-based polymers functionalized with azobenzene dyes. Molecular weight determination of these polymers was performed using gel permeation chromatography. The weight average molecular weights of the ferrocene polymers were in the range of 8 000-12 000 while the molecular weights of the cationic-neutral organoiron polymers were estimated using these values to be in range of 11 000-16 000.

Thermal gravimetric analysis showed that the cleavage of the cationic cyclopentadienyliron moieties occurred between 230 and 249 °C, while the polymeric backbone degraded between 424 and 465 °C.

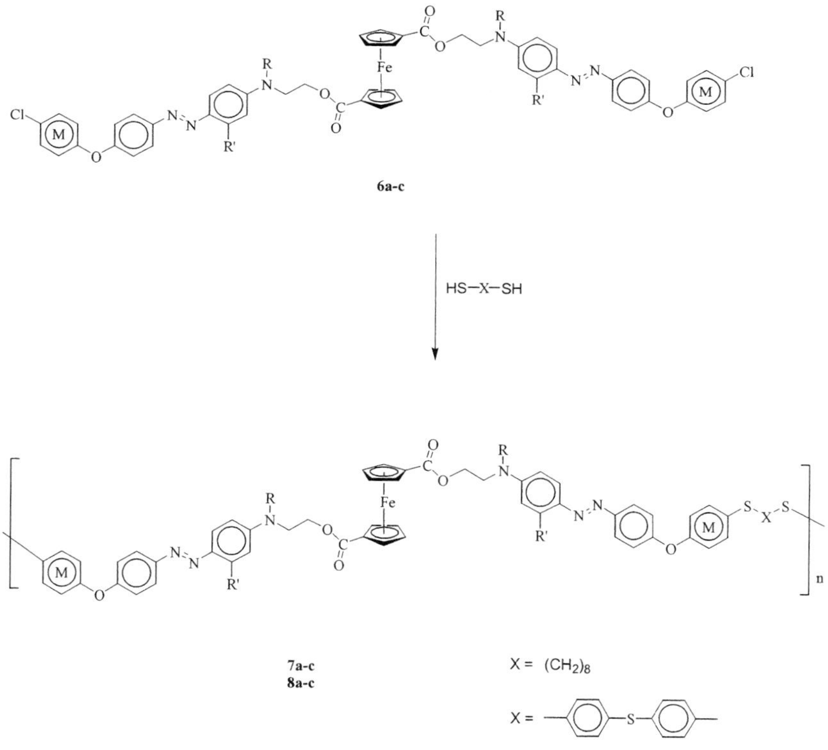

Scheme 3

UV-vis analysis illustrated that the monomeric and polymeric materials display similar wavelength maxima, which was attributed to the π-π* and n- π* transitions of the incorporated azo dyes. The UV-vis studies were carried out in DMF and DMF/HCl solutions to examine the halochromisim of these polymers. Figure 2 shows the visible spectra of polymer **7b**, which displays a λ_{max} at 433 nm in DMF solution. Addition of 10% HCl to the polymer solution produced a bathochromic shift (λ_{max} at 525 nm). This bathocrhomic shift has been attributed to the formation of the azonium ion.

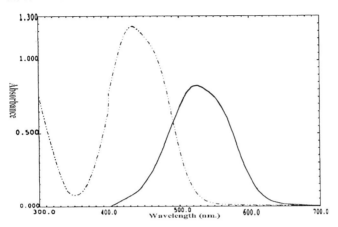

Figure 2. Visible spectra of polymer **7b** in DMF and DMF/HCl solution.

Conclusions

Organoiron macromolecules containing azo dyes in the backbones or pendant to the side chains were synthesized under very mild reaction conditions using a metal-mediated methodology. Thermal analysis of the metallated polymers showed two weight losses corresponding to the cleavage of the cationic moieties and the degradation of the polymer backbones. Design of trimetallic complexes with terminal CpFe$^+$ and inner bis(cyclopentadienyliron) moieties was accomplished and their polymerization afforded coloured macromolecules containing neutral and cationic organoiron centers. It is also important to note that photolytic demetallation of these polymeric materials resulted in the cleavege of CpFe$^+$ and therefore the isolation of novel ferrocene-based polymers with azo dyes in the backbone.

Acknowledgements

Financial support provided by the Natural Sciences and Research Council of Canada (NSERC) and Manitoba Hydro are gratefully acknowledged. R.M.O. would also like to thank the Department of Chemistry, University of Manitoba.

[1]   A. S. Abd-El-Aziz, *Macromol. Rapid Commun.* **2002**, *23*, 995.
[2]   A. S. Abd-El-Aziz, E.K. Todd, *Coord. Chem. Rev.* **2003**, *246*, 3.
[3]   R. D. A. Hudson, *J. Organomet. Chem.* **2001**, *47*, 637.
[4]   R. D. Archer, in *"Inorganic and OrganometallicPolymers"*, Wiley-VCH, New York 2001.
[5]   A. S. Abd-El-Aziz, C.E. Carraher Jr., C. U. Pittman Jr., J. E. Sheats, M. Zeldin, *"Macromolecules Containing Metal and Metal-Like Elements"*, volume 1, Wiley & Sons Inc., New Jersy 2003.
[6]   A. S. Abd-El-Aziz, C.E. Carraher Jr., C. U. Pittman Jr., J. E. Sheats, M. Zeldin,*"Macromolecules Containing Metal and Metal-Like Elements"*, volume 2, Wiley & Sons Inc., New Jersy 2003.
[7]   A. S. Abd-El-Aziz, "Metal-Containing Polymers" in: *Encyclopedia of Polymer Science and Technology*, 3rd Edition, J. I. Kroschwitz, Ed., John Wiley & Sons, New York **2002**.
[8]   F. S. Arimto, A. C. Haven Jr., *J. Am. Chem. Soc.* **1955**, *81*, 6295.
[9]   C. U. Pittman Jr., C. E. Carraher Jr., J. R. Reynolds, "Organometallic Polymers", in: *Encyclopedia of Polymer Science and Technology*, J. I. Kroschwitz, H. F. Mark, N. M. Bikales, C. G. Overberger, G. Menges, Eds., John Wiley & Sons, New York 1987.
[10] P. Nguyen, P. Gomez-Elipe, I. Manners, *Chem. Rev.* **1999**, *99*, 1515.
[11] D. Astruc, *"Electron Transfer and Radical Processes in Transition-Metal Chemistry"*, VCH publishers Inc., New York 1995.
[12] A. S. Abd-El-Aziz, S. Bernardin, *Coord. Chem. Rev.* **2000**, *203*, 219.
[13] A. S. Abd-El-Aziz, E. K. Todd, G. Z. Ma, *J. Polym. Sci., Part A: Polym. Chem.* **2001**, *39*, 1216.
[14] A. S. Abd-El-Aziz, E. K. Todd, T. H. Afifi, *Macrmol. Rapid Commun.* **2002**, *23*, 113.
[15] A. S. Abd-El-Aziz, E. K. Todd, R. M. Okasha, T. E. Wood, *Macromol. Rapid Commun.* **2002**, *23*,743.
[16] A. S. Abd-El-Aziz, L. J. May, J. A. Hurd, R. M. Okasha, *J. Polym. Sci., Part A: Polym. Chem.* **2001**, *39*, 2716.
[17] A. Natansohn, P. Rochon, *Chem. Rev.* **2002**, 102, 4139.
[18] M. Hasegawa, T. Ikawa, M. Tsuchimori, O. Watanabe, *J. Appl. Polym. Sci.* **2002**, *86*, 17.
[19] L. A. Howe, G. D. Jaycox, *J. Polym. Sci., Part A: Polym. Chem.* **1998**, *36*, 2827.
[20] K. Huang, H. Qiu, M. Wan, *Macromolecules* **2002**, *35*, 8653.
[21] S. H. Kang, H.-D. Shin, C. H. Oh, D. H. Choi, K. H. Park, *Bull. Korean Chem. Soc.* **2002**, *23*, 957.
[22] P. Uznanski, J. Pecherz, *J. Appl. Polym. Sci.* **2002**, 86, 1456.
[23] C. Samyn, T. Verbiest, A. Persoons, *Macromol. Rapid Commun.* **2000**, *21*, 1.
[24] G. Iftime, F. L. Labarthet, A. Natansohn, P. Rochon, K. Murti, *Chem. Mater.* **2002**, *14*, 168.
[25] G. H. Kim, C. D. Keunm, S. J. Kim, L. S. Park, *J. Polym. Sci., Part A: Polym. Chem.* **1999**, 37, 3715.
[26] K. Tawa, K. Kamada, K. Kiyohara, K. Ohta, D. Yasumatsu, Z. Sekkat, S. Kawata, *Macromolecules* **2001**, *34*, 8232.
[27] L. Wu, X. Tuo, H. Cheng, Z. Chen X. Wang, *Macromolecules* **2001**, *34*, 8005.
[28] X.-H. Liu, D. W. Bruce, I. Manners, *Chem. Commun.* **1997**, 289.

Macromol. Symp. **2004**, *209*, 195-205

Organoiron Polynorbornenes with Pendent Arylazo and Hetarylazo Dye Moieties

Alaa S. Abd-El-Aziz,* Rawda M.Okasha, Tarek H. Afifi

Department of Chemistry, The University of Winnipeg, Winnipeg, Manitoba, Canada
R3B 2E9
E-mail: a.abdelaziz@uwinnipeg.ca

Summary: Organoiron polynorbornene containing arylazo or hetarylazo dye chromophores has been prepared via ring opening metathesis polymerization using Grubbs' catalyst. The obtained polymers were isolated as brightly colored materials and displayed good solubility in polar organic solvents. The colors of these polymers were affected by the nature of the incorporated azo chromophores. Thermogravimetric analysis of these materials showed that the cleavage of the cyclopentadienyliron (CpFe$^+$) moieties was between 225 and 231 °C, while the degradation of the polymer backbones occurred between 400 and 450 °C. UV-vis studies in DMF showed that the organoiron polymers containing arylazo dyes exhibit wavelength maxima around 425 nm. However, the replacement of these arylazo moieties with hetarylazo dyes displayed substantial bathochromic shifts in the λ_{max} values ($\approx$ 511 nm).

Keywords: arene complexes; arylazo dyes; cyclopentadienyliron; hetarylazo dyes; polynorbornene

Introduction

Ring opening metathesis polymerizations (ROMP) of cycloalkenes and bicycloalkenes have been extensively explored.[1-16] The development of highly active metal carbene catalysts allows for controlling the molecular weight and the backbone configurations of the resulting polymers.[3-8] To date, norbornene remains one of the most studied molecules due to its facile functionalizations and high reactivity in ROMP. [1-16] The introduction of various substituents to norbornene monomers and polymers allows for varying the physical and the chemical properties

© 2004 WILEY-VCH Verlag GmbH & KGaA, Weinheim   DOI: 10.1002/masy.200450514

of these materials. Some of the properties include liquid crystallinty, redox, luminescence as well as biological properties. [9-16]

The past few decades have seen tremendous attention given to the development of metal-containing polymers. This interest stems from the unique properties that this class of polymers possess. For example, the incorporation of metal moieties into the macromolecules backbone and side chains enhances the solubility of these materials and offers interesting properties such as hardness, magnetism, catalytic behavior and redox activity. [17-29]

Our research has focused on the design of novel iron-based macromolecules using the cyclopentadienyliron methodology. The exceptional electron-withdrawing ability of the cationic iron moieties facilitates the nucleophilic aromatic substitution reactions of the chloroarene complexes and allows for the isolation of novel monomeric and polymeric material under mild conditions.[17, 20-22, 26-32] Recently we have reported the synthesis of thermally stable polynorbornene containing pendant etheric linkages.[30-32] Polynorbornenes functionalized with organometallic moieties in their side chains have also been investigated. Albagli and coworkers[11] have reported the synthesis of electrochemically-active polynorbornenes containing ferrocenyl moieties in their side chains. We also synthesized the first examples of polynorbornenes substituted with side chains containing CpFe$^+$ complexes.[30]

Currently, we are exploring the incorporation of arylazo moieties into macromolecules backbones and side chains. Polymers containing azo dyes are potential candidates for reversible optical storage systems, electrooptic modulators, and photorefractive switches.[33-40] In one of our recent articles, we reported the first example of cationic iron-coordinated polyaromatic ethers and thioethers functionalized with azo dyes in their side chains.[41] As well, we have also reported ring opening metathesis polymerization of novel organoiron norbornene monomers containing arylazo moieties using Grubbs catalyst.[42] Addition polymerization of norbornene monomers functionalized with arylazo chromophores have also been reported.[39]

This article is focused on the introduction of hetarylazo dyes into the norbornene monomers. The colour of the resulting organometallic polymers containing these dyes ranged from reddish-violet to blue depending on the nature of the dye moiety. Herein, we will describe the synthetic strategies, spectroscopic and analytical characterizations of the monomeric and polymeric materials.

Results and Discussion

Monomer Synthesis

Using metal-mediated nucleophilic aromatic substitution reactions, various monometallic complexes containing arylazo moieties with terminal hydroxyl groups have been prepared. Scheme 1 shows the design of organoiron norbornene monomers functionalized with the arylazo chromophores. Reactions of chloroarene complexes with phenolic arylazo molecules resulted in the formation of organoiron arylazo complexes containing terminal hydroxyl groups (**2a,b-4a,b**). These complexes were subsequently reacted with *exo-*, *endo*-5-norbornene-2-carboxylic acid (**1**) resulting in the formation of novel norbornene monomers functionalized with cyclopentadienyliron-dye complexes **5a,b-7a,b**.

Characterizations of these monomeric species were determined using one- and two-dimensional NMR spectroscopy due to the existence of the *exo-* and *endo*-isomeric structures. For example, Figure 1 shows the HH COSY NMR of the aliphatic protons of monomer **5a**. Identification of *endo* and *exo* isomers was determined based on the assignment of the olefinic protons $H_6(n)$ and $H_5(n)$, which are strongly coupled with each other and resonate between 5.86 and 6.15 ppm respectively. The bridgehead protons $H_1(n)$ (3.21 ppm) and $H_4(n)$ (2.81ppm) appeared as broad singlets and identified via their coupling with $H_6(n)$ and $H_5(n)$. The connectivities between $H_4(n)$ and the H_3 protons as well as the strong coupling between $H_3(x)$ and $H_3(n)$ protons led to the assignment of $H_3(x)$ to the multiplet at 1.89-1.93 ppm and $H_3(n)$ to the multiplet at 1.27-1.41 ppm. The $H_{7a}(n)$ and $H_{7s}(n)$ were also strongly coupled to each other and resonated within the multiplet at 1.27 to 1.41 ppm and the doublet at 1.50 ppm. The remaining norbornene protons were determined using the same methodology.

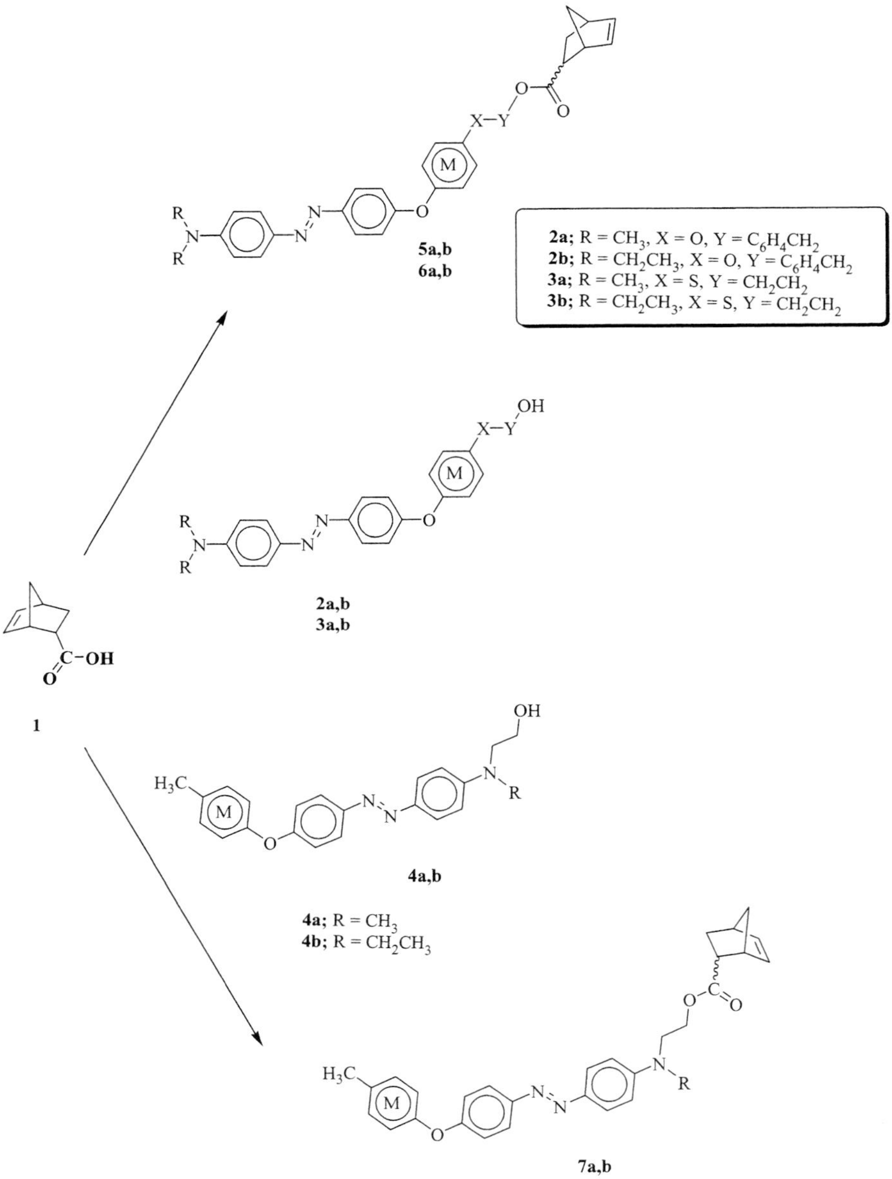

2a; R = CH$_3$, X = O, Y = C$_6$H$_4$CH$_2$
2b; R = CH$_2$CH$_3$, X = O, Y = C$_6$H$_4$CH$_2$
3a; R = CH$_3$, X = S, Y = CH$_2$CH$_2$
3b; R = CH$_2$CH$_3$, X = S, Y = CH$_2$CH$_2$

5a,b
6a,b

2a,b
3a,b

1

4a,b

4a; R = CH$_3$
4b; R = CH$_2$CH$_3$

7a,b

Scheme 1

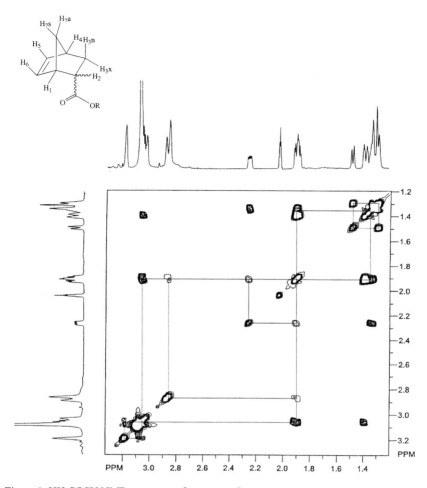

Figure 1. HH COSY NMR spectrum of monomer **5a**.

Condensation reactions of cyclopentadienyliron complexes containing carboxylic groups with hydroxyl groups benzothiazole azo dyes allowed for the isolation of novel organoiron complexes incorporating hetarylazo dyes. These complexes were reacted with 4-hydroxy benzyl alcohol and then capped with the norbornene molecules to form the first example of organoiron norbornene monomers functionalized with hetarylazo dye chromophores as shown in Scheme 2. The monomeric units were isolated in good yields, and identified using spectroscopic analysis.

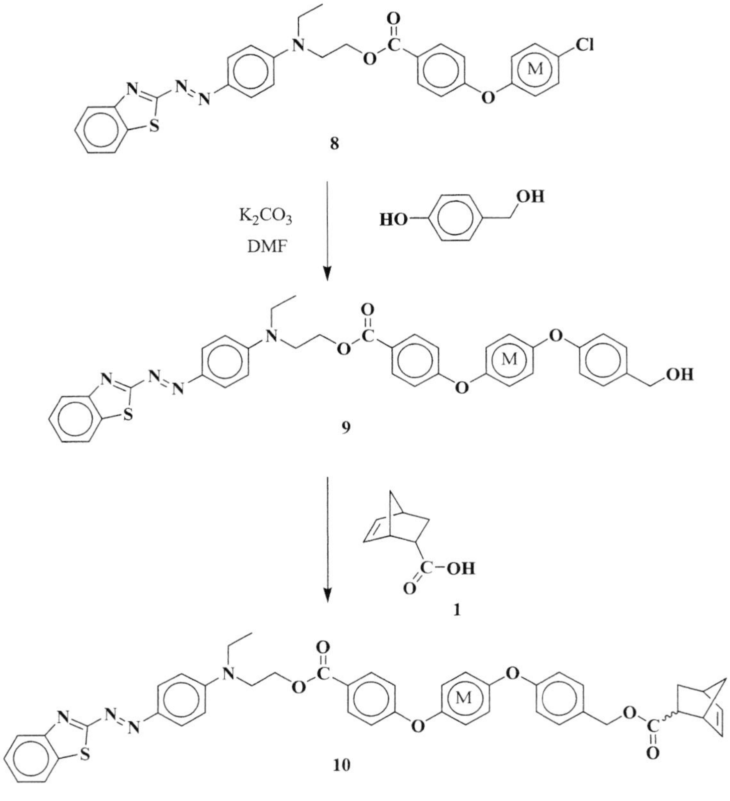

Scheme 2

Polymerization

Ring opening metathesis polymerization of norbornene monomers (**5a,b, 6a,b, 7a,b, 10**) using Grubbs' catalyst with monomer/initiator ratio [20]:[1] led to the isolation of metallated polynorbornenes containing azo dyes in the side chains. The obtained polymers displayed excellent solubility in DMF, DMSO and acetonitrile. While polymers **11a,b-13a,b** were orange solids, polymer **14** was isolated as purple solid. This change in color was attributed to the incorporation of benzothiazole azo chromophore into the side chain of polynorbornene.

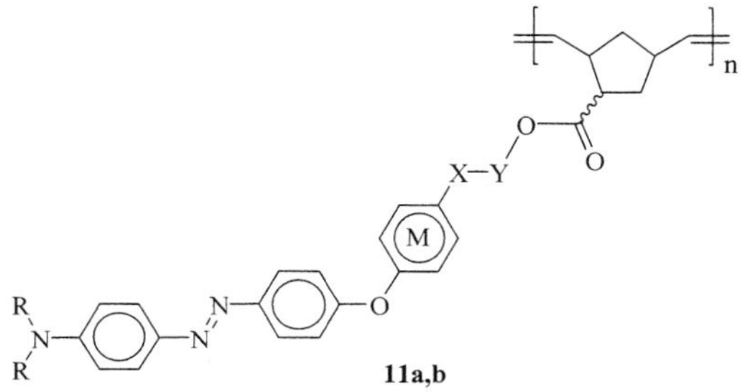

11a,b
12a,b

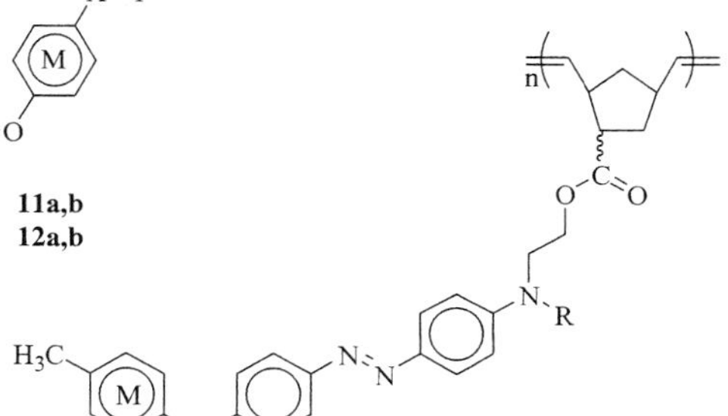

13a,b

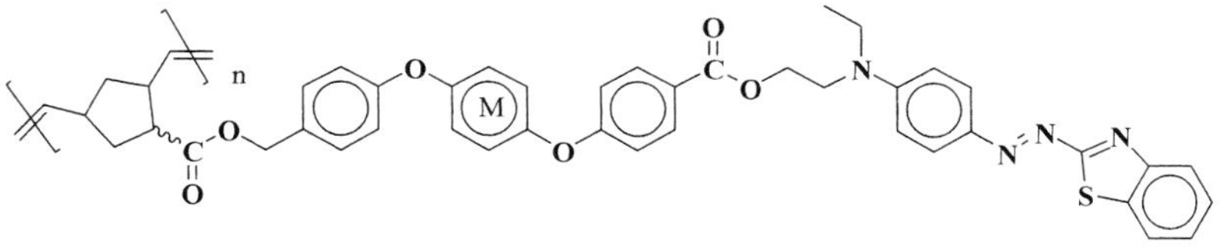

14

Characterizations of the polymeric materials were performed using spectroscopic and thermal techniques. As an example, the ^{1}H NMR spectra of the resulting polymers showed disappearance of the olefinic protons from the norbornene monomer between 5.78 and 6.18 ppm, and the presence of a new set of peaks between 5.00 and 5.37 ppm, which is consistent with the double bond formation of the polymer backbone and thus verified the success of the polymerization.

Photolytic demetallation of the organoiron polymers was performed prior to the molecular weight determination due to the interaction of the cyclopentadienyliron moieties with the gel permeation chromatography (GPC) columns. The molecular weights of the organic polymers were determined using GPC and were in the range of 9 200-21 800 which corresponds to M_w of 14 300-31 600 for the metallated analogues. Thermal stability of the metallated and demetallated functionalized polymers was examined using thermogravimetric analysis (TGA) and differential scanning calorimetry (DSC). The TGA data of the metallated polynorbornene showed that the cationic organoiron moieties cleave between 225 and 231 °C, while the degradation of the polymer backbones occurred between 400 and 450 °C. Differential scanning calorimetric studies illustrated that the glass transition temperatures (T_g) of the resulting polymers was affected by the length of the side chains, the position of the cationic iron moieties and the substituted azo chromophores. The T_g's values for the metallated polymers were in the range of 140-178 °C while their organic analogues exhibit T_g's between 75 and 104 °C. Polymer **12b** shows a T_g at 140 °C (Figure 2).

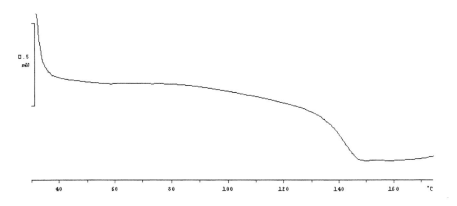

Figure 2. DSC thermogram of polymer **12b**.

UV-visible spectroscopic analysis of the colored organoiron polynorbornene was examined. The wavelength maxima of the obtained polymers were influenced by the nature of the azo chromophore moieties. For example, the incorporation of benzothiazole based chromophore instead of an arylazo dye into the polymer side chain displayed substantial bathochromic shifts in the λ_{max} values. This strong bathochromic effect may be attributed to the d-orbitals of the sulfur. The λ_{max} of polymers **13a** in DMF occurs at 424 nm, while polymer **14** exhibits wavelength maximum at 511 nm (Figure 3, curve **a** and **b** respectively). These values increased after the addition of a hydrochloric acid solution due to the formation of the azonium ions.[43] Polymers **13a** show λ_{max} at 520 nm while the maximum of polymer **14** occurred at 608 nm with a shoulder at 574 nm, which is consistent with the π-π* transition of azo dyes (Figure 3).

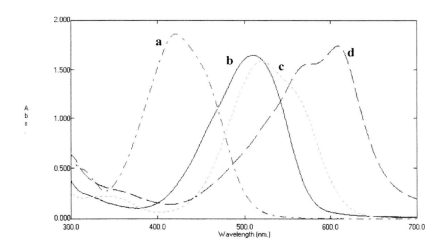

Figure 3. UV-vis spectra of (a) polymer **13a** in DMF, (b) polymer **14** in DMF, (c) polymer **13a** in DMF/HCl and (d) polymer **14** in DMF/HCl.

Conclusions

Novel organoiron polynorbornene functionalized with arylazo and hetarylazo dyes have been synthesized. The molecular weight of the metallated polymers was estimated to be in the range of 14 300-31 600. Differential scanning calorimetry showed that the T_g's of the organoiron polymers ranged from 140 to 178 °C, while their organic analogues displayed T_g values from 75 to 104 °C. Visible spectra of the resulting colored polymers in DMF and DMF/HCl solutions showed that the nature of the incorporated azo chromophore influences the wavelength maxima of these materials.

Acknowledgements

Financial support provided by the Natural Sciences and Research Council of Canada (NSERC) and Manitoba Hydro are gratefully acknowledged. R.M.O. would also like to thank the Department of Chemistry, University of Manitoba.

[1] K. J. Ivin, J. C. Mol, "*Olefin Metathesis and Metathesis Polymerization*", Academic Press, San Diego 1997.
[2] V. Dragutan, R. Streck, "*Catalytic Polymerization of Cycloolefins: Ionic, Ziegler-Natta and Ring-Opening Metathesis Polymerization*", Elsevier, Amsterdam, 2000.
[3] M. Scholl, S. Ding, C. W. Lee, R. H. Grubbs, *Org. Lett.* **1999**, *1*, 953.
[4] P. Schwab, R. H. Grubbs, J. W. Ziller, *J. Am. Chem. Soc.* **1996**, *118,* 100.
[5] R. R. Schrock, J. S. Murdzek, G. C. Bazan, J. Robbins, M. DiMare, M. O'Regan, *J. Am. Chem. Soc.* **1990**, *112*, 3875.
[6] T. M. Trnka, R. H. Grubbs, *Acc. Chem. Res.* **2001**, *34*, 18.
[7] R. R. Schrock, *Chem. Rev.* **2002**, *102*, 145.
[8] M. R. Buchmeiser, *Chem. Rev.* **2000**, *100*, 1565.
[9] G.-H. Kim, C. Pugh, S. Z.-D. Cheng, *Macromolecules* **2000**, *33*, 8983.
[10] Z. Liu, L. Zhu, Z. Shen, W. Zhou, S. Z. D. Cheng, V. Percec, G. Ungar, *Macromolecules* **2002**, *35*, 9246.
[11] D. Albagli, G. C. Bazan, R. R. Schrock, M. S. Wrighton, *J. Am. Chem. Soc.* **1993**, *115*, 7328.
[12] K. J. Watson, J. Zhu, S. T. Nguyen, C. A. Mirken, *J. Am. Chem. Soc.* **1999**, *121*, 462.
[13] T. J. Boyd, R. R. Schrock, *Macromolecules* **1999**, *32*, 6608.
[14] T. J. Boyd, Y. Geerts, J.-K. Lee, D. E. Fogg, G. G. Lavoie, R. R. Schrock, M. F. Rubner, *Macromolecules* **1997**, *30*, 3553.
[15] H. D. Maynard, S. Y. Okada, R. H. Grubbs, *J. Am. Chem. Soc.* **2001**, *123*, 1275.
[16] S. Franceschi, O. Bordeau, C. Millerioux, E. Perez, P. Vicendo, I. Rico-Lattes, A. Moisand, *Langmuir* **2002**, *18,* 1743.
[17] A. S. Abd-El-Aziz, E. K. Todd, *Coord. Chem. Rev.* **2002**, *246*, 3.
[18] R. D. A. Hudson, *J. Organomet. Chem.* **2001**, *47*, 637.
[19] R.D. Archer, in "*Inorganic and Organometallic Polymers*'', Wiley-VCH, New York 2001.
[20] A. S. Abd-El-Aziz, C. E. Carraher Jr., C. U. Pittman Jr., J. E. Sheats, M. Zeldin, "*Macromolecules Containing Metal and Metal-Like Elements*", Volume 1, Wiley & Sons Inc., New Jersy 2003.
[21] A. S. Abd-El-Aziz, C. E. Carraher Jr., C. U. Pittman Jr., J. E. Sheats, M. Zeldin,"*Macromolecules Containing Metal and Metal-Like Elements*", Volume 2, Wiley & Sons Inc., New Jersy 2003.

[22] A. S. Abd-El-Aziz, "Metal-Containing Polymers" in: *Encyclopedia of Polymer Science and Technology*, 3rd Edition, J. I. Kroschwitz, Ed., John Wiley & Sons, New York 2002.

[23] T. Masuda, T. Higashimura, *Adv. Polym. Sci.* **1987**, *81*, 121.

[24] C. U. Pittman Jr., C. E. Carraher Jr., J. R. Reynolds, "Organometallic Polymers'', in: *Encyclopedia of Polymer Science and Technology*, J. I. Kroschwitz, H. F. Mark, N. M. Bikales, C. G. Overberger, G. Menges, Eds., John Wiley & Sons, New York 1987.

[25] P. Nguyen, P. Gomez-Elipe, I. Manners, *Chem. Rev.* **1999**, *99*, 1515.

[26] A. S. Abd-El-Aziz, C. R. de Denus, K. M. Epp, S. Smith, R. J. Jaeger, D. T. Pierce, *Can. J. Chem.* **1996**, *74*, 650.

[27] A. S. Abd-El-Aziz, E. K. Todd, R. M. Okasha, T. E. Wood, *Macromol. Rapid Commun.* **2002**, *23*, 743.

[28] A. S. Abd-El-Aziz, E. K. Todd, G. Z. Ma, *J. Polym. Sci., Part A: Polym. Chem.* **2001**, *39*, 1216.

[29] A. S. Abd-El-Aziz, E. K. Todd, G. Z. Ma, J. DiMartino, *J. Inorg. Organomet. Polym.* **2000**, *10*, 265.

[30] A. S. Abd-El-Aziz, L. J. May, J. A. Hurd, R. M. Okasha, *J. Polym. Sci., Part A: Polym. Chem.* **2001**, *39*, 2716.

[31] A. S. Abd-El-Aziz, L. May, A. L. Edel, *Macromol. Rapid Commun.* **2000**, *21*, 598.

[32] A. S. Abd-El-Aziz, A. L. Edel, L. May, K. M. Epp, H. M. Hutton, *Can. J. Chem.* **1999**, *77*, 1797.

[33] A. Natansohn, P. Rochon, *Can. J. Chem.* **2001**, *79*, 1093.

[34] A. Natansohn, P. Rochon, *Chem. Rev.* **2002**, *102*, 4139.

[35] G. S. Kumar, D. C. Neckers, *Chem. Rev.* **1989**, *89*, 1915.

[36] M. Hasegawa, T. Ikawa, M. Tsuchimori, O. Watanabe, *J. Appl. Polym. Sci.* **2002**, *86*, 17.

[37] L. A. Howe, G. D. Jaycox, *J. Polym. Sci.: Part A: Polym. Chem.* **1998**, *36*, 2827.

[38] K. Huang, H. Qiu, M. Wan, *Macromolecules* **2002**, *35*, 8653.

[39] S. H. Kang, H.-D. Shin, C. H. Oh, D. H. Choi, K. H. Park, *Bull. Korean Chem. Soc.* **2002**, *23*, 957.

[40] H. Zollinger, "*Diazo Chemistry I*", VCH Publishers Inc., New York 1994.

[41] A. S. Abd-El-Aziz, T. H Afifi. W. R. Budakowski, K. J. Friesen, E. K. Todd, *Macromolecules* **2002**, *35*, 8929.

[42] A. S. Abd-El-Aziz, R. M. Okasha , T. H Afifi, E. K. Todd, *Macromol. Chem. Phys.* **2003**, *204*, 555.

[43] P. Uznanski, J. Pecherz, *J. Appl. Polym. Sci.* **2002**, *86*, 1456.

Synthesis of Sulfone-Containing Monomers and Polymers Using Cationic Cyclopentadienyliron Complexes

Alaa S. Abd-El-Aziz,* Nelson M. Pereira, Waleed Boraie, Shaune L. McFarlane, Erin K. Todd

Department of Chemistry, The University of Winnipeg, Winnipeg, Manitoba, Canada R3B 2E9
E-mail: a.abdelaziz@uwinnipeg.ca

Summary: The synthesis of sulfone-containing monomers with pendent cationic cyclopentadienyliron ($CpFe^+$) moieties has been accomplished via nucleophilic aromatic substitution of dichloroarene complexes with a number aliphatic dithiols. These complexes were further oxidized using m-CPBA to give the sulfone-based monomers. Polymerization of the sulfone-based monomers with O-containing nucleophiles produced the sulfone-based polymers. Direct nucleophilic aromatic substitution of dichloroarene complexes with dinucleophiles allowed for the formation of organoiron sulfide-based polymers. Oxidation of these polymers led to the formation of sulfone polymers with the pendent iron moieties. The organometallic monomers and polymers were found to be more soluble in polar solvents in comparison to their organic analogues.

Keywords: arene complexes; cyclopentadienyliron; organoiron polymers; polyether and thioethers; sulfur- and sulfone-based polymers

Introduction

There has been a tremendous interest in the development of sulfur-containing polymers due to their wide spread industrial applications as electrical components and in the manufacturing of car engine parts.[1] Sulfone-based polymers such as poly(aromatic ether sulfone)s are known as engineering thermoplastics.[2-5] This class of macromolecules is known for its excellent thermal stability and mechanical properties. These polymers also have good hydrolytic stability, outstanding toughness, low smoke emission, and electrical insulating properties.[2, 6] Introduction of pendent groups to the sulfone-based polymers could lead to improved transport properties.[6]

There are a number of traditional organic routes to the synthesis of sulfone-containing monomers and polymers.[7] An alternative method is using arenes coordinated to the cyclopentadienyliron

© 2004 WILEY-VCH Verlag GmbH & KGaA, Weinheim   DOI: 10.1002/masy.200450515

moiety.[8] Complexation of chloroarenes to cyclopentadienyliron cations increases the reactivity of the arene towards nucleophiles, therefore decreasing reaction times and reducing extremely harsh conditions. For example, reaction of a dichlorobenzene with ferrocene in the presence of a Lewis acid produces the organoiron complex, which can then be reacted with various thiols or thiophenols to produce the corresponding sulfide-functionalized complexes.[9-11] Oxidation of the sulfide groups produces sulfones, which are excellent activating groups for nucleophilic aromatic substitution reactions. Polymerization of the sulfone-substituted chloroarenes can be achieved directly, or following demetallation of the organoiron complexes.

Results and Discussion

Over the past decade, we have been interested in the development of synthetic routes to organometallic as well as organic monomers and polymers using cationic cyclopentadienyliron ($CpFe^+$) and pentamethylcyclopentadienylruthenium (Cp^*Ru^+) complexes.[12-18] In this study, two strategies were developed for the preparation of sulfone-based polymers. The first strategy involved the synthesis of diiron sulfide complexes via metal-mediated aromatic nucleophilic substitution reactions. Oxidation of these complexes using m-CPBA allowed for the isolation of their corresponding sulfones. The complexes were then demetallated using pyrolysis to give the organic sulfones. These sulfone monomers were then polymerized to produce the sulfone-based polymers. The second strategy entailed the direct polymerization of the dichlorobenzene with ethane-, butane-, hexane-, and octane-dithiol. The resulting polymeric sulfides were then oxidized to give the organoiron sulfone-based polymers.

The bimetallic aliphatic sulfide complexes (**3a-d**) were prepared via the nucleophilic aromatic substitution reactions of (dichlorobenzene)$CpFe^+$ complexes with dithiols in DMF in the presence of K_2CO_3 as shown in Scheme 1. Oxidation of the diiron sulfide complex **3a-d** gave rise to the sulfone complex **4a-d** in good yields.

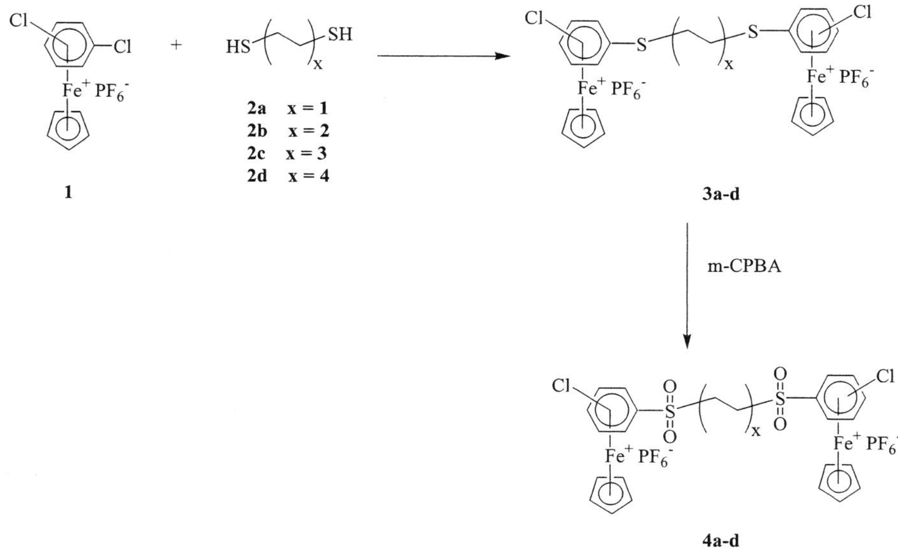

Scheme 1

The following ^{1}H-NMR spectra (Figure 1) are of the bimetallic sulfide **3b** and bimetallic sulfone complex **4b**. The strong electron-withdrawing capabilities of the sulfones caused a downfield shift of the cyclopentadienyl and complexed aromatic resonances in the NMR spectra compared to the corresponding sulfide complexes.

Pyrolysis of the organometallic monomers **4a-d** resulted in the isolation of their corresponding organic analogues **5a-d** (Scheme 2). These monomers were then characterized by ^{1}H- and ^{13}C-NMR as well as IR spectroscopy. Figure 2 shows the ^{1}H-NMR of monomer **5b**. The cyclopentadienyliron peak is no longer there in the spectrum and the formerly complexed aromatic protons are shifted downfield in the spectrum of the organic sulfone monomer. The two peaks at approximately 3.0 ppm and 2.8 ppm are the two CH$_2$ groups, with the peak resonating more downfield being the one bonded α to the sulfone and the other peak, being the methyl group β to the sulfone group.

210

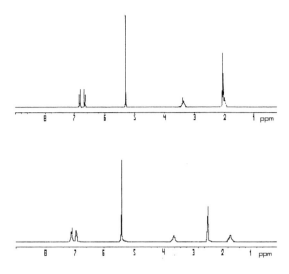

Figure 1. ¹H-NMR spectra of sulfide, 3b (top, acetone-d₆) and sulfone, 4b (bottom, DMSO-d₆)

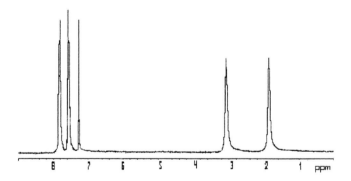

Figure 2. ¹H- NMR of demetallated sulfone monomer 5b (CDCl₃)

Demetallation of **4b,** as an example, was achieved by first grinding it into a fine powder and placing it in the pyrolytic chamber where it was heated to 250 °C under reduced pressure for 2 h. After purification through a silica column, the solution was then concentrated and precipitated into hexane affording the sulfone monomer **5b**. The synthesis of **6a-c** was performed in DMF and K_2CO_3 under a nitrogen atmosphere, where the temperature was maintained between 155-165 °C for 16 h. The solution was then precipitated in 10% HCl filtered and washed with water.

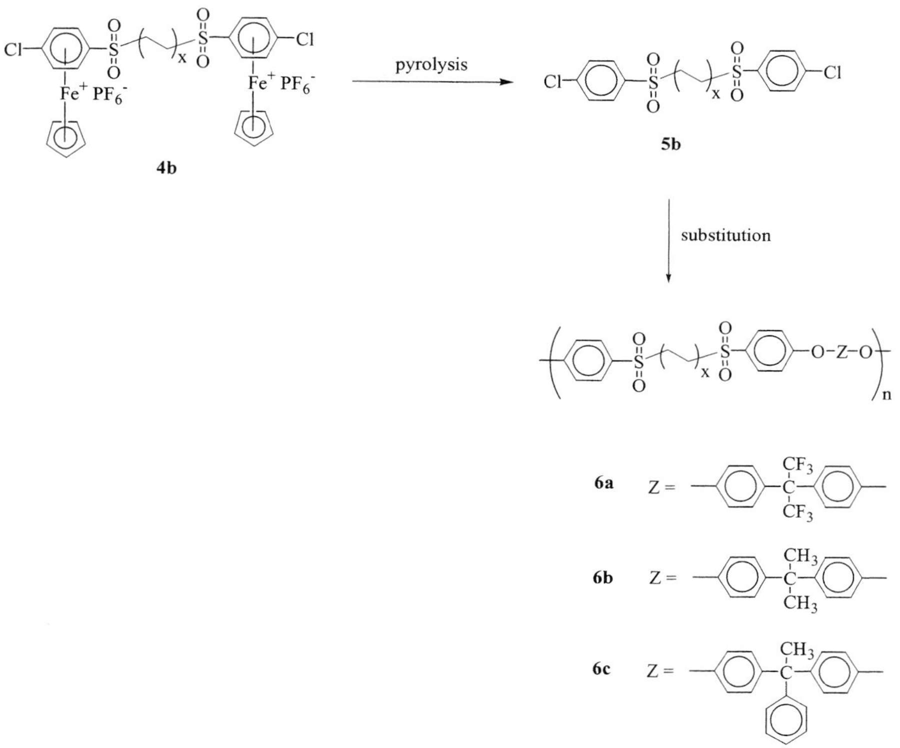

Scheme 2

The successful polymerization of the monomer **5b** with bisphenol A was determined by the upfield shift of the resonances of the protons that are adjacent to the chloro groups. The protons adjacent or α to the sulfone resonate at 2.94 ppm, the β protons appeared at 1.59 ppm, and the methyl groups on the bisphenol A appeared at 1.70 ppm. The aromatic protons appeared at 6.84-6.97 ppm (m, 8H), 7.16 (d, 4H, J = 8.2 Hz), 7.67 (d, 4H, J = 8.2 Hz). The APT (attached proton test) ^{13}C-NMR is further proof that the polymerization occurred, as shown by the quaternary ArC peaks appearing at 132.02, 147.26, 152.61, and 162.62 ppm. The ArCH peaks appear at 117.61, 119.86, 128.47, and 129.63 ppm. The α CH_2 peak appears at 55.58 ppm, the β CH_2 appears at 21.58 ppm; the methyl group resonates at 30.9 ppm; finally the quaternary carbon appears at 42.38 ppm.

Infrared analysis was also performed to confirm the presence of the sulfone and the ether groups. The sulfone group should show stretching in the range of 1 160-1 120 cm^{-1} and the C-O-C should stretch at around 1 230-1 250 cm^{-1}. The IR shows that the sulfone appears at 1 145 cm^{-1} and the ether at 1 245 cm^{-1}. Gel permeation chromatography was used to determine that the M_w was 9 100 with a polydispersity of 2.03, indicating that the degree of polymerization was 16. TGA showed that the onset and endset of decomposition for this polymer occur at 421 and 482 °C, respectively, while DSC analysis showed that the T_g was 142 °C.

The influence of aliphatic chains in the backbones of polythioethers was analysed by reacting complex **1** with **2** as shown in Scheme 3.[13] Polymers **7a-c** were isolated in 88-91% yield as beige coloured precipitates.

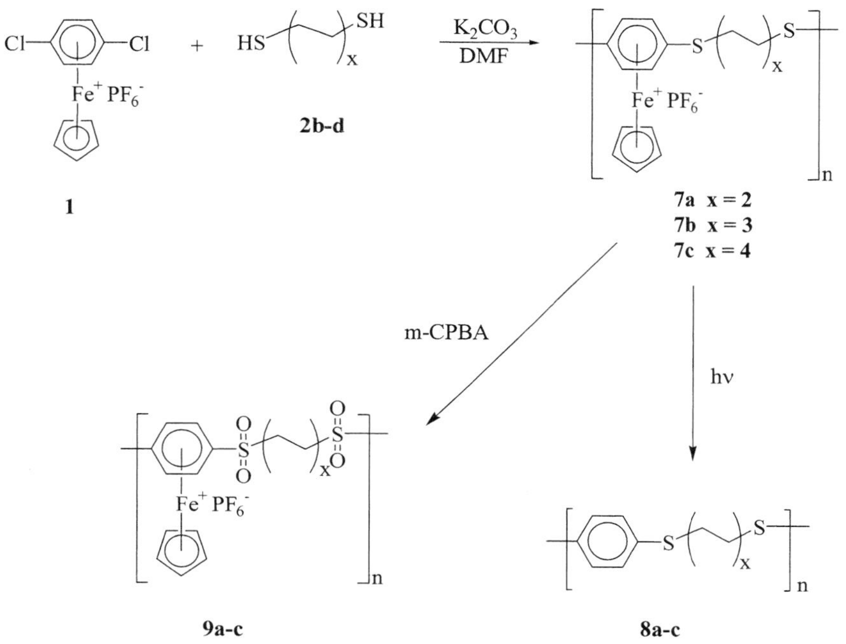

Scheme 3

Polymer **7a** displayed very low solubility in various polar solvents and precipitated from DMF during the polymerization process. It was observed that with the increase in the size of the aliphatic chains in the polymer backbones, there was a corresponding increase in polymer solubility. As an example, the polymer that incorporated the octamethylene groups in the backbone was formed more rapidly, was more soluble, and had a higher molecular weight than polymers with shorter aliphatic chains in the polymer backbone. As well, the 1,6-hexanedithiol (**7b**) and 1,8-octanedithiol (**7c**) polymers were much more stable at high temperatures during the polymerization reactions, which was most likely due to their higher solubilities in the reaction medium.

Figure 3 shows the ^{1}H NMR spectrum of polymer **7b**. There are three resonances corresponding to the methylene peaks in the polymer backbone at 1.42, 1.71, and 3.26 ppm. The cyclopentadienyl protons appear as a singlet at 5.01 ppm, while the complexed aromatic protons appear as a singlet at 6.41 ppm. After demetallation, the aromatic peak in **8b** shifted downfield and appeared as a singlet at 7.20 ppm.

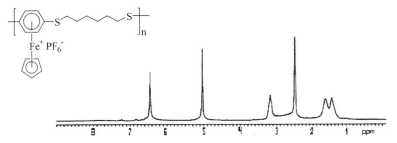

Figure 3. ^{1}H NMR spectrum of polymer 7b in DMSO-d$_6$

When oxidation of **7a** was performed, the ^{1}H-NMR of both **7a** and **9a** were compared to confirm that the oxidation was successful. For the Cp peak, it shifted downfield from 5.01 ppm in the sulfide polymer to 5.45 ppm in the sulfone polymer. The aromatic peaks appear at 6.46 ppm and 7.33 ppm for the sulfide and sulfone polymers, respectively. The α CH$_2$ peak resonates at 3.45 ppm in the sulfide polymer and at 3.71 ppm in the sulfone polymer. Lastly, the β CH$_2$ group resonates at 1.70 ppm and at 1.75 ppm for the sulfide and sulfone polymers, respectively.

The solubilities of polymer **8a-c** were also highly dependent on the number of methylene units in their backbones. It was found that polymers **8b** and **8c** were soluble in organic solvents such as THF and chloroform due to the flexible six-and eight-carbon aliphatic chains in their backbones. These polymers were determined to have weight average molecular weights of 13 500 and 21 400, respectively. Due to the short aliphatic spacer in polymer **8a**, this polymer was found to be only partially soluble. The TGA thermograms of **8a-c** were as follows: for **8a** the start of decomposition occurred at 344 °C with the end of decomposition at 400 °C; for **8b** the onset was at 372 °C and the endset at 428 °C; finally for **8c** the onset was at 327 °C and the endset occurred at 447 °C. The T$_g$s of **8a-c** was 37, 34, and 33 °C, respectively.

The cyclic voltammogram of polymer **7c** is shown in Figure 4. At a scan rate of 5 V/s, this CV shows the two sequential one-electron reduction steps of the iron centres pendent to the polymer backbone. It was found that at low scan rates, the second reduction step was irreversible. The $E_{1/2}$ values corresponding to formation of the neutral nineteen-electron, and anionic twenty-electron iron species occurring at -1.07 and -1.77 V, respectively.

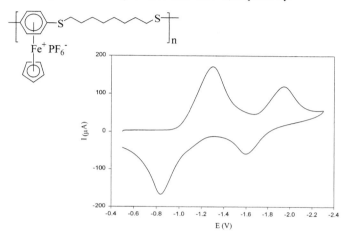

Figure 4. Cyclic voltammogram of 7c obtained at a scan rate of 5 V/s

Conclusion

The synthesis of sulfone-based polymers can be successfully conducted via metal-mediated methodology. Two different routes were followed in order to achieve this goal. The first route involved the synthesis of disulfide monomers, that were oxidized, followed by pyrolytic demetallation to give the sulfone-based monomers, which were then polymerized. The second route involved the direct polymerization of a dichlorobenzene complex with a number of aliphatic dithiols. Oxidation of these polymers with m-CPBA produced the corresponding organometallic polysulfones. The length of the aliphatic chains in the backbone showed a direct correlation of the solubilities of these polymers. While the hexane and octane spacers in the polymers enhanced the solubilities, the incorporation of butane spacers gave rise to partially insoluble polymers.

Acknowledgements

Financial support provided by the Natural Sciences and Research Council of Canada (NSERC), Manitoba Hydro and the University of Winnipeg are gratefully acknowledged.

[1]   E. S. Wilks, Ed., „*Industrial Polymers Handbook*", Wiley-VCH, Weinheim 2001, p.1243-1282.
[2]   A. H. Frazer, „*High Temperature Resistent Polymers*", John Wiley and Sons, Toronto 1968.
[3]   V. Percec, R. S. Clough, M. Grigoras, P. L. Rinaldi, V. E. Litman, *Macromolecules* **1993**, *26*, 3650.
[4]   V. Percec, R. S. Clough, M. Grigoras, P. L. Rinaldi, V.E. Litman, *Macromolecules* **1994**, *27*, 1535.
[5]   J. Lu, K. Miyatake, A. R. Hill, A. S. Hay, *Macromolecules* **2001**, *34*, 5860.
[6]   R. J. Cotter, „*Engineering Plastics, A Handbook of Polyarylethers*", Gordon and Breach Publishers, Basel 1995.
[7]   H. R. Allcock, F. W. Lampe. „*Contemporary Polymer Chemistry*", Prentice Hall, New Jersey 1990, p.41-42.
[8]   A. S. Abd-El-Aziz, S. Bernardin, *Coord. Chem. Rev.* **2000**, *203,* 21.
[9]   A. S. Abd-El-Aziz, K. M. Epp, Y. Lei, S. Kotowich, *J. Chem. Research* (S) **1995**, 182.
[10] A. S. Abd-El-Aziz, K. M. Epp, C. R. de Denus, G. Fisher-Smith, *Organometallics* **1994**, *13*, 2299.
[11] A. S. Abd-El-Aziz, S. L. M\ :sup:`c`\Farlane, T. H. Afifi, T. C. Corkery, *Am. Chem. Soc. Div. Polym. Chem.* **2002**, *43(1),* 506.
[12] C. R. de Denus, L. M. Hoffa, E. K. Todd, A. S. Abd-El-Aziz, *J. Inorg. Organomet. Polym,* **2000**, *10*, 189.
[13] A. S. Abd-El-Aziz, E. K. Todd, G. Z. Ma, *J. Polym. Sci. Part A: Polym. Chem.* **2001**, *39*, 1216.
[14] A. S. Abd-El-Aziz, E. K. Todd, R. M. Okasha, T. E. Wood, *Macromol. Rapid Commun.* **2002**, *23*, 743.
[15] A. S. Abd-El-Aziz, E. K. Todd, *Coord. Chem. Rev.* **2003**, *246*, 3.
[16] A. S. Abd-El-Aziz, T. C. Corkery, E. K. Todd, T. H. Afifi, G. Z. Ma, *J. Inorg. Organomet. Polym.* **2003**, *13*, 113.
[17] A. S. Abd-El-Aziz, R. M. Okasha, T. H. Afifi, E. K. Todd, *Macromol. Chem. Phys.* **2003**, *204*, 555.
[18] A. S. Abd-El-Aziz, E. K. Todd, R. M. Okasha, T. E. Wood, *Macromol. Rapid Commun.* **2002**, *23*, 743.

Synthesis and Applications of Polymers Containing Metallacyclopentadiene Moieties in the Main Chain

Ikuyoshi Tomita,* Masahiro Ueda

Department of Electronic Chemistry, Interdisciplinary Graduate School of Science and Engineering, Tokyo Institute of Technology, Nagatsuta-cho 4259, Midori-ku, Yokohama 226-8502, Japan
Email: tomita@echem.titech.ac.jp

Summary: Organometallic polymers containing metallacycles in the main chain were prepared by the reactions of diynes with low-valent organometallic complexes such as $CpCo(PPh_3)_2$, $CP_2Ti(CH_2=CHC_2H_5)$, and $(^iPrO)_2Ti(CH_2=CHCH_3)$. Their polymer reactions involving the conversion of the main chain structures gave rise to polymers containing functional groups in their main chain repeating units. Design and synthesis of organometallic polymers that potentially serve as novel functional materials are also described.

Keywords: metallacycles; metallocenes; organometallic polymers; polymer reactions; reactive polymers

Introduction

Polymers containing reactive organometallic components in the main chain are potentially useful to create new reactive polymers that can yield organic polymers with versatile main chain functionality. Incorporation of organometallic cores with unique geometry and/or characters such as electrochemical properties may also be useful to design functional materials. Accordingly, we have been working on the synthesis of polymers with organometallic repeating units such as cobaltacyclopentadiene and titanacyclopentadiene moieties in the main chain. This article covers the results obtained recently in our group related to the synthesis and applications of organometallic polymers.

Synthesis of **Cobaltacyclopentadiene-Containing Polymers**

According to Yamazaki and his co-workers, cobaltacyclopentadiene derivatives are readily accessible by the reaction of $CpCo(PPh_3)_2$ (**1**) and acetylene derivatives.[1] On the basis of the air-stability and moderate reactivity of the cobaltacyclopentadiene derivatives, polymers containing the corresponding units are attractive to realize novel kinds of reactive and functional materials. Accordingly, the polymerization of CpCo(PPh$_3$)$_2$ (**1**) and diynes (**2**) was performed in toluene at 50-60 °C, from which polymers containing cobaltacyclopentadiene moieties in the main chain (**3**) were obtained in high yields (Scheme 1).[2] The brown-colored polymers (**3**) thus obtained are soluble in organic solvents and are stable under air, although they are occasionally contaminated with cyclobutadienecobalt moieties (~10%). In terms of the regiochemistry of the main chain connections at each cobaltacyclopentadiene unit, it is possible to form three regioisomeric units (i.e., the connections through 2,5-, 2,4-, and 3,4-positions of the metallacycles) which can be controlled to some extent by the substituents on the diyne monomers. That is, diynes with less sterically hindered lateral substituents such as **2b** and **2c** gave rise to polymers with a higher content of the 2,5-linkage (**3b**: ~70% and **3c**: ~100%). However, the solubility of the polymers in organic solvents decreased as the 2,5-content increased.

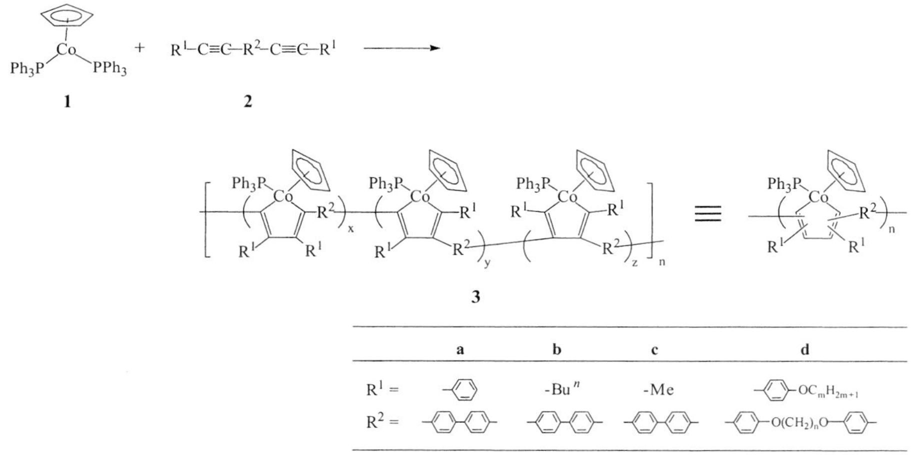

Scheme 1

In accordance with the decomposition temperature of the cobaltacyclopentadiene derivatives (e.g., 193-194 °C for a tetraphenyl substituted cobaltacyclopentadiene[1a]), the organocobalt polymers (**3**) were found to start decomposing at approximately 200 °C by the thermogravimetric analyses (TGA) and two-step decomposition was observed in all the organocobalt polymers. The first decomposition at about 200 °C can be attributed to the rearrangement of the cobaltacyclopentadiene moieties into the cyclobutadienecobalt which accompanies the elimination of triphenylphosphine (vide infra) and the second one above 400 °C to the decomposition of the organic fragments. No peak for either the glass transition (T_g) nor the melting temperature (T_m) is observable for **3a** in its differential scanning calorimetric (DSC) analysis. Using diynes with flexible spacers (**2d**), it is possible to design organocobalt polymers (**3d**) which exhibit T_g and T_m.[3] The polymers (**3d**) have T_g at -20~130 °C, depending upon the length of the aliphatic spacers. Some of the polymers also exhibit T_m in the range of 60~120 °C.

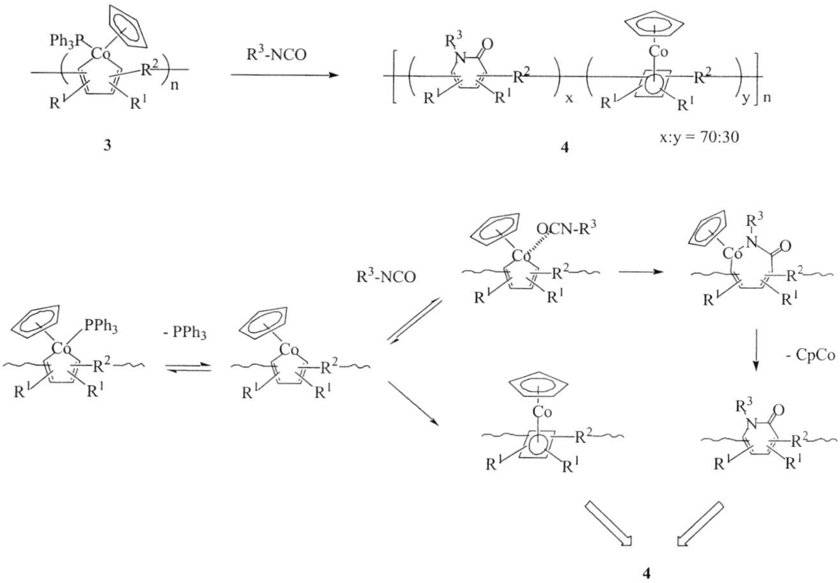

Scheme 2

Conversion into Organic Polymers with Various Main Chain Structures

As mentioned above, the organocobalt polymers are stable under air. However, they serve as novel type of reactive polymers whose main chain can be reconstructed by polymer reactions under appropriate conditions. Conversion of the metallacycle units in the organocobalt polymers

(**3**) into organic functional groups is attainable by the polymer reactions with appropriate reagents. As shown in Scheme 2, polymers having pyridone moieties in the main chain (**4**) were produced from **3** by the reaction with isocyanates at 120 °C.[4] The content of the 2-pyridone moieties reaches about 70% with respect to the starting cobaltacyclopentadiene units. The remaining 30% was found to be η^4-cyclobutadienecobalt moieties as a result of the rearrangement reaction.

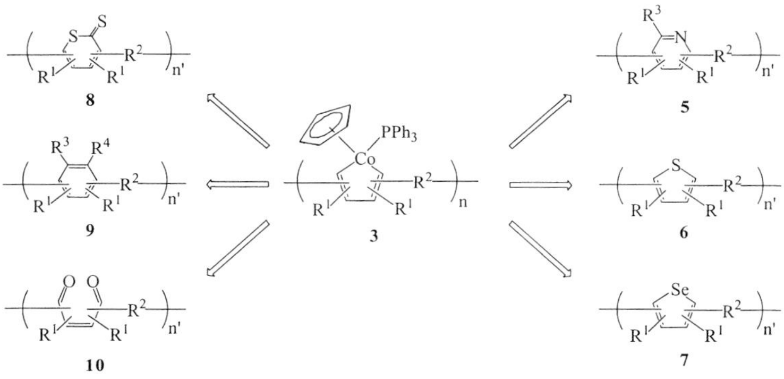

Scheme 3

As summarized in Scheme 3, the organocobalt polymers can be converted into various kinds of polymers with versatile functional groups in the main chain. That is, polymers containing pyridine (**5**), thiophene (**6**), selenophene (**7**), dithiolactone (**8**), phenylene (**9**), and diketone moieties (**10**) were obtained by the reaction with nitriles, sulfur, selenium, carbon disulfide, acetylenes, and oxygen, respectively.[5-10] The organic polymers derived from the organocobalt polymers with fully aromatic main chain systems (e.g., the thiophene-containing polymers from **3a-c**) exhibited the properties of π-conjugated oligomers judging from their UV-vis spectra and the electrochemical properties, probably due to the regio-irregularity of the main chain connection of the organocobalt polymers. The efficiency of these polymer reactions is affected by the reagents and the reaction conditions, which ranged from 70-100%. Even in the cases of polymer reactions that do not proceed in a quantitative fashion, no distinct decrease of the molecular weight of the produced polymers was observed because the lower efficiency of the polymer reactions does not mean the scission of the main chain of the polymers but the conversion into other structures such as the cyclobutadienecobalt unit.

Conversion into Other Organocobalt Polymers

On heating the organocobalt polymers (3) in the presence of appropriate ligands such as trialkylphosphines, triphenylphosphine on the organocobalt polymers (3) can be replaced quantitatively by the added ligands, by which the properties such as the solubility of the polymers can be modified (Scheme 4).[11]

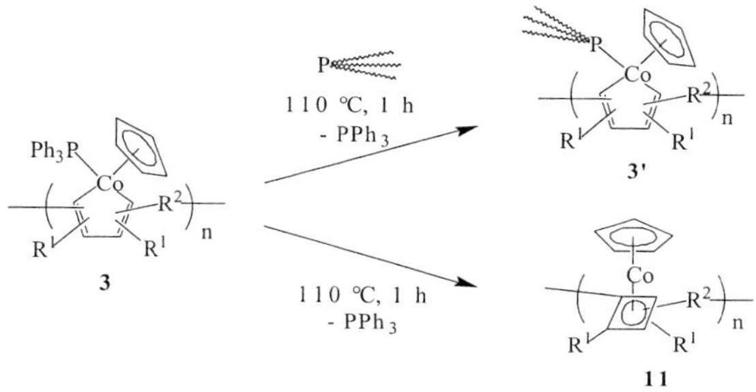

Scheme 4

Without the addition of ligands, the thermal treatment of the polymers (3) gave rise to yellow-colored polymers (11) containing η^4-cyclobutadienecobalt moieties as a result of the thermal rearrangement reaction.[12] This reaction seems to proceed by the dissociation of the ligand followed by the elimination of the cobalt. The cyclobutadienecobalt-containing polymers (11) are also stable under air and soluble in organic solvents such as chloroform, THF, and N,N-dimethylformamide (DMF). The polymer (11a) produced from 3a exhibited good thermal stability and the 5% weight loss was observed at 480 °C in its TGA. The cobaltacyclopentadiene-containing polymers with flexible spaces (3d) also give rise to cyclobutadienecobalt-containing polymers (11d) by the thermal rearrangement. The resulting cyclobutadienecobalt-containing polymers (11d) also have T_g and T_m.[3]

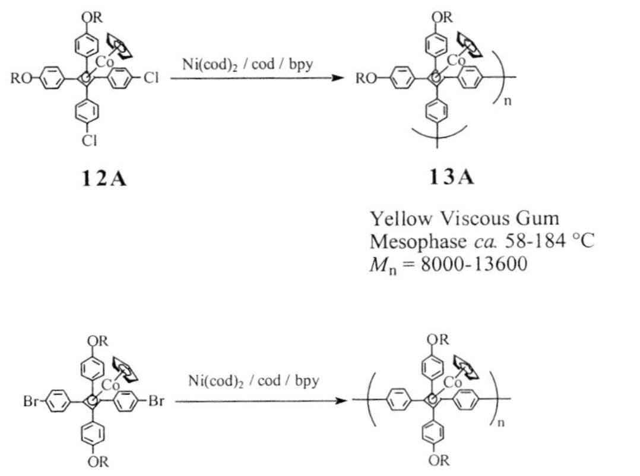

12A → **13A**

Yellow Viscous Gum
Mesophase *ca.* 58-184 °C
M_n = 8000-13600

12B → **13B**

$R = -C_6H_{13}{}^n \sim -C_{16}H_{33}{}^n$

Yellow Powder (partly soluble)
Mesophase *ca.* 156-280 °C
M_n = 16700-22000

Scheme 5

Unique functional materials containing cyclobutadienecobalt moieties might be designed on the basis of their chemical and thermal stability, and of the peculiar square geometry of the cyclobutadiene ligands. Reflecting upon the squarer geometry of the cyclobutadiene moieties, thermotropic liquid crystalline polymers with regioregular main chain connections can be designed by the polycondensation of the regioisomerically pure organocobalt monomers.[13-16] For example, the Ni(0)-promoted dehalogenation polycondensation of both 1,2- and 1,3-regioisomerically pure bifunctional $(\eta^5$-cyclopentadienyl)$(\eta^4$-cylobutadiene)cobalt complexes (**12A** and **12B**) yields zigzag and rigid rod π-conjugated polyarylenes (**13A** and **13B**), respectively (Scheme 5).[15] Both of these polymers exhibit thermotropic liquid-crystalline behavior, where the zigzag type polymers (**13A**) exhibit thermotropic liquid crystals in lower temperature range and higher solubility in organic solvents compared to the rigid rod type polymers (**13B**). Using a mixture of **12A** and **12B** as a monomer for the polycondensation, the liquid crystalline properties of the polymers could be varied by the ratio of the two isomers.[17] Similar to the cases of the conversion of the organocobalt polymers (**3**) into organic polymers, **3** can also be converted into other organometallic polymers by the reactions with appropriate reagents. Polymers having $(\eta^5$-cyclopentadienyl)$(\eta^4$-iminocyclopentadiene)cobalt moieties in the main chain (**14**) were obtained by the reaction with isocyanides (Scheme 6).[18]   It is of note that the efficiency of the polymer reaction was quantitative and the polymers produced exhibit a unique solvatochromism.   That is, a polymer solution exhibits a reversible color change from purple to red by varying the nature of the solvent (e.g., purple in benzene and red in methanol). This color change might be ascribable to the structural change between the neutral (η^4-iminocyclopentadiene)cobalt (**14**) and the zwitterionic cobalticenium unit (**14'**).   The subsequent polymer reaction of **14** with alkyl halides gives polymers containing cobalticenium units (**15**) although the efficiency of the alkylation was not quantitative due to the precipitation of the polymer during the reaction.   The electrochemical analysis of **15** suggests the presence of the electronic interaction between the plural organometallic centers.

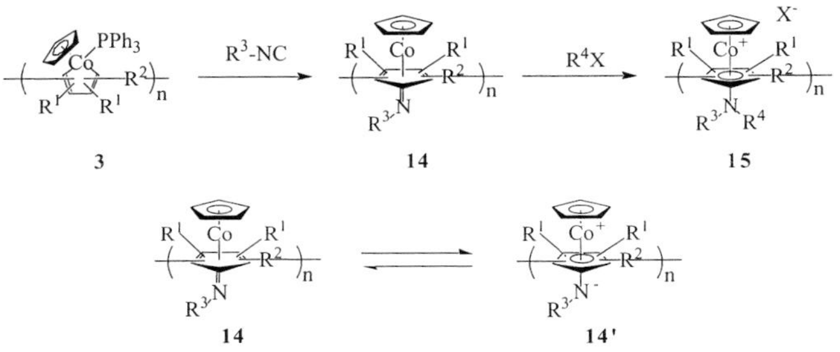

Scheme 6

It is reported that the reaction of cobaltacyclopentadiene derivatives with carbon disulfide gives unsaturated dithiolactones in moderate yields[1k] by which dithiolactone-containing polymers (**8**) can be produced from **3** as described above.   On the contrary, the addition of an equimolar amount of Co(I) complexes such as CpCo(cod) to the reaction system provided an entirely different result.   In this case, the dithiolactones were not detected at all but (η^4-cyclopentadiene)cobalt complexes are obtained in excellent yields.[19]   On the basis of this reaction, analogous cobalticenium-containing polymers (**16**) are also obtainable by the reaction of the polymers (**3**) with carbon disulfide in the presence of a Co(I) complex, followed by the S-alkylation (Scheme 7).[20]

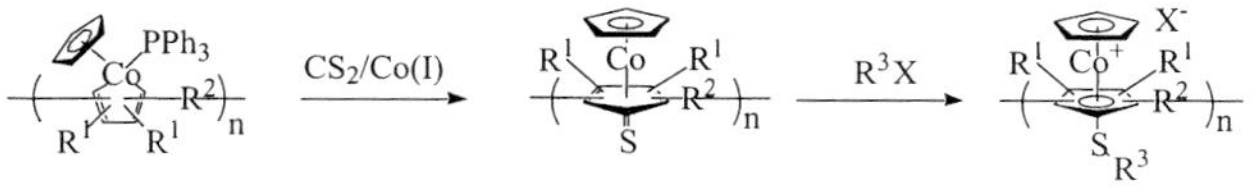

Scheme 7

Synthesis and Reactions of Titanacyclopentadiene-Containing Polymers

Polymers containing other metallacycle units are attractive for creation of novel reactive polymers leading to versatile organic functional polymers. As described by Tilley et al., zirconacyclopentadiene-containing polymers can be prepared by means of the analogous metallacyclization process between a low-valent bis(cyclopentadienyl)zirconium and diynes. The resulting zirconium-containing polymers also reveal interesting reactivity giving rise to organic polymers diene and heterocyclic moieties in the main chain.[21] Owing to the recent progress of the chemistry of titanacycles,[22] polymers possessing titanacyclopentadiene units are also potentially attractive for the creation of main chain reactive materials. On the basis of the titanacycle formation process described recently by Takahashi et al.,[22a] it is possible to prepare titanacyclopentadiene-containing polymers from diynes (2) and a low-valent titanocene derivative (17) generated from bis(cyclopentadienyl)titanium dichloride (Scheme 8). The resulting polymers (18) are stable at ambient temperature under argon atmosphere.

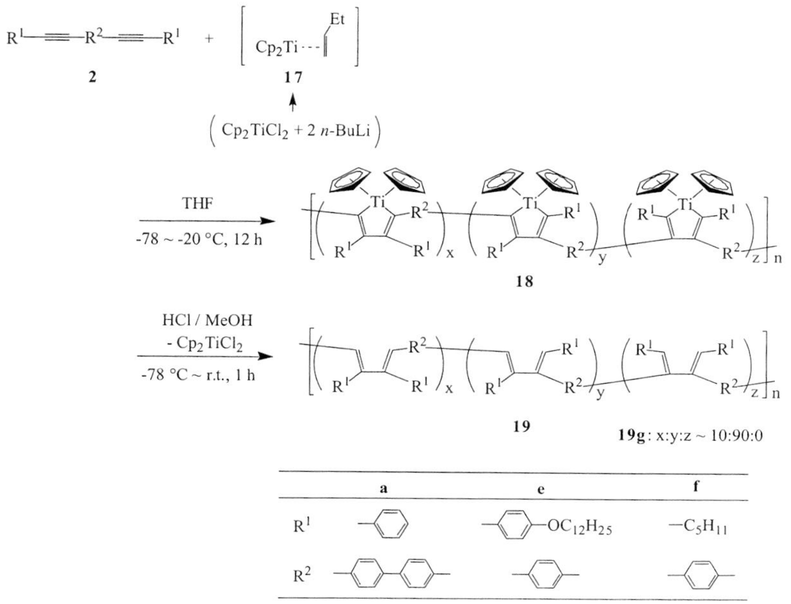

Scheme 8

The hydrolytic work-up of the polymers (**18**) gives rise to diene-containing polymers (**19**) in moderate yields ($M_n \sim 4000$). The regioisomeric linkage of the main chain of **18** at the titanacycle moieties must be dependent upon the diynes used and the polymers (**18a** and **18e**) may have statistical distribution of 2,5-, 2,4-, and 3,4-linkages although these units are difficult to be distinguished by the spectroscopic methods. In the case of the polymer (**18f**), the regiochemistry could be determined by the model experiment. That is, the titanacycles obtained from 1-heptynylbenzene, two phenyl substituents are located at the 2,4- and the 2,5-positions (2,4-:2,5- = 90:10) judging from their hydrolysis products. Accordingly, the main chain of the polymer (**18f**) is supposed to be connected through the 2,4- and the 2,5- positions of the metallacycle moieties. Because of the rather low content of the 2,5-connection, an effective linkage for π-conjugation, the polymer (**19f**) produced exhibited a relatively small red shift in the UV-vis spectrum in comparison with the model dienes. The result can also be taken to mean that the polymer (**18f**) produced from the low-valent titanocene derivative and diyne (**2f**) has the major 2,4- and the minor 2,5- connections through the titanacycle units.

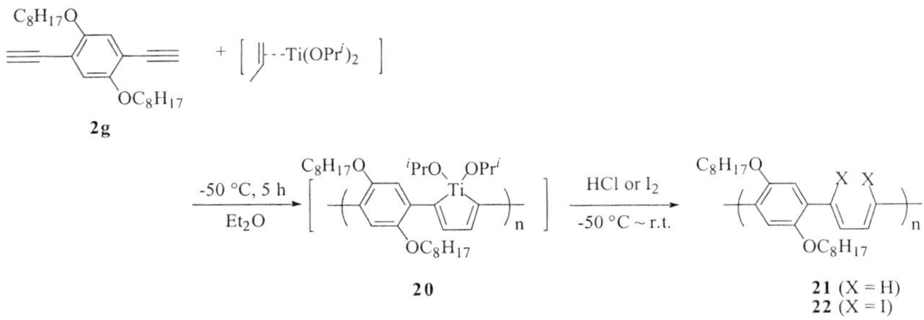

Scheme 9

As shown in Scheme 9, titanacyclopentadiene-containing polymers with a regiospecific main chain connection could be obtained by the polymerization of terminal diynes (**2g**) and a low-valent titanium generated from titanium(IV) isopropoxide.[24]   The polymerization proceeds at −78 ~ −50 °C and the polymers produced should be converted into organic polymers without isolation because they are not stable at ambient temperature.   According to the report of Yamaguchi et al.,[22b] 1,4-disubstituted dienes are produced in the reaction of terminal acetylenes such as phenylacetylene followed by the hydrolysis.   In accordance with their report, the polymerization of terminal diynes such as 1,4-diethynyl-2,5-dioctyloxybenzene (**2g**) gave rise to polymers with regioregular backbone (**20**).   The hydrolysis or the iodination of the titanacycle units gave rise to polymers with diene units (**21** and **22**, respectively).   For example, the treatment of the polymer (**20**) with iodine provided an iodinated diene-containing polymer (**22**) in 81% yield (M_n = 7,700).   Owing to the alkoxy substituents, both the polymers (**21** and **22**) are soluble in organic solvents.   The regioregularity of the main chain affects largely on the properties of the diene-containing polymers.   The polymer (**21**) produced by the hydrolysis of the titanacyclopentadiene-containing polymer exhibits a clear bathochromic shift of the UV-vis absorption compared to that of a model compound (**23**) (Figure 1).   That is, the absorption maximum (λ_{max}) of the polymer (λ_{max} = 470 nm) appeared at longer wavelength by 115 nm than that of the model compound (λ_{max} = 355 nm). The polymers also exhibit luminescence upon irradiation of the UV-vis light.

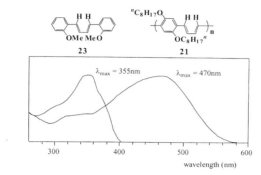

Figure 1. UV-vis spectra of the diene-containing polymer (**21**) and its model compound (**23**) (measured in CHCl$_3$).

Derivatives of poly(p-phenylene) (**24**) are produced by the reaction of the titanacyclopentadiene-containing polymers (**20**) with propargyl bromide (Scheme 10).[25] Similar to the case of the chemical conversion into diene-containing polymers, the polymer reaction proceeds under very mild conditions and yellow powdery polymers produced are soluble in organic solvents (~ 80% yield, M_n~ 5,000). Because of the π-conjugated backbone, the poly(p-phenylene) derivatives (**24**) obtained in this study have the UV-vis absorption in longer wavelength range (λ_{max} = 329 nm) compared to that of model compounds (λ_{max} = 276 nm), p-terphenyl derivatives produced from phenylacetylene derivatives via the titanacyclopentadiene. The polymers also exhibited photoluminescence whose emission maximum also shifts to the longer wavelength region with respect to that of the model compounds.

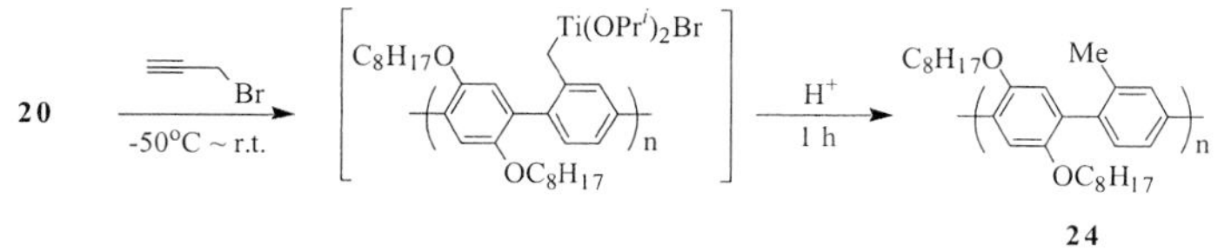

Scheme 10

Thiophene-containing polymers (**25**) are likewise produced by the reaction of the titanacyclopentadiene-containing polymers (**20**) with sulfur monochloride under mild conditions (Scheme 11).[26]   The yield of the soluble fraction was a little lower (~ 50%), because the produced polymers are partially insoluble in organic solvents.   As shown in Figure 2, the brown powdery polymers thus obtained exhibit a substantial bathochromic shift in the UV-vis and photoluminescence spectra in comparison with those of the corresponding model compounds, a 2,5-diarylthiophene derivative (**26**).

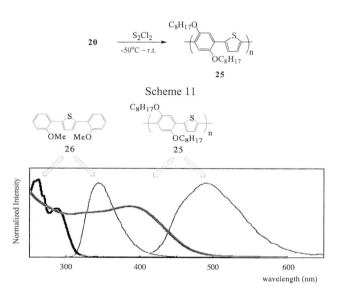

Scheme 11

Figure 2. UV-vis (—) and photoluminescence (—) spectra of the thiophene-containing polymer (**25**) and its model compound (**26**) (measured in CHCl$_3$).

The titanacyclopentadiene-containing polymers (**20**) can be converted into π-conjugated phosphole-containing polymers (**27**) by the reaction with dichlorophenylphosphine (Scheme 12).[27]   Similar to the case of the above-mentioned conversion reaction into thiophene-containing polymers, the produced polymers are partially insoluble in organic solvents.   The polymers thus obtained revealed λ$_{max}$ at about 500 nm in the UV-vis spectra and the emission maximum at 600 nm, supporting the π-conjugated character of the resulting phosphole-containing polymers (Figure 3).

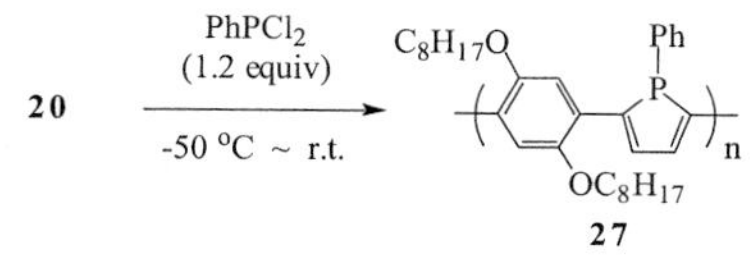

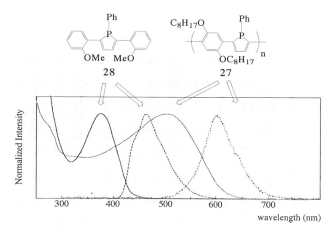

Scheme 12

Figure 3. UV-vis (——) and photoluminescence (····) spectra of the phosphole-containing polymer (**27**) and its model compound (**28**) (measured in CHCl₃).

Conclusion

Novel polymers possessing reactive sites in the main chain have been successfully prepared by the reactions of diynes with low-valent transition metal complexes such as cobalt(I) and titanium(II). The cobaltacyclopentadiene-containing polymers are stable under air and can be handled as if they are conventional organic polymers. However, they exhibit versatile reactivity under appropriate reaction conditions. That is, the organocobalt polymers serve as reactive precursors for a variety of polymers containing functional groups such as heterocycles, benzene rings, and unsaturated ketones in their repeating unit. The organocobalt polymers can also be

230

converted to other organocobalt polymers with unique repeating units such as cyclobutadienecobalt and cobalticenium moieties. The titanacyclopentadiene-containing polymers are likewise obtained by the metallacyclization process of diynes with low-valent titanium complexes. Although the two series of the titanacyclopentadiene-containing polymers are not stable under air, they also serve as reactive precursors giving rise to organic polymers with versatile main chain structures.

[1] (a) H. Yamazaki, Y. Wakatsuki, *J. Organomet. Chem.* **1977**, *139*, 157. (b) Y. Wakatsuki, O. Nomura, K. Kitaura, K. Morokuma, H. Yamazaki *J. Am. Chem. Soc.* **1983**, *105*, 1907. (c) Y. Wakatsuki, H. Yamazaki, *J. Chem. Soc., Dalton Trans.* **1982**, 1923. (d) P. Hong, H. Yamazaki, *Synthesis* **1977**, 50. (e) K. Yasufuku, A. Hamada, K. Aoki, H. Yamazaki, *J. Am. Chem. Soc.* **1980**, *102*, 436. (f) H. Yamazaki, N. Hagihara, *J. Organomet. Chem.* **1967**, *7*, P22. (g) Y. Wakatsuki, H. Yamazaki, *Tetrahedron Lett.* **1973**, 3383. (h) Y. Wakatsuki, H. Yamazaki, *Synthesis* **1976**, 26. (i) Y. Wakatsuki, H. Yamazaki, *J. Chem. Soc., Dalton Trans.* **1978**, 1278. (j) H. Yamazaki, Y. Wakatsuki, *Bull. Chem. Soc. Jpn.* **1979**, *52*, 1239. (k) Y. Wakatsuki, H. Yamazaki, *J. Chem. Soc., Chem. Commun.* **1973**, 280.

[2] (a) I. Tomita, A. Nishio, T. Igarashi, T. Endo, *Polym. Bull.* **1993**, *30*, 179 (b) J.-C. Lee, A. Nishio, I. Tomita, T. Endo, *Macromolecules* **1997**, *30*, 5205.

[3] I. L. Rozhanskii, I. Tomita, T. Endo, *Macromolecules* **1996**, *29*, 1934.

[4] I. Tomita, A. Nishio, T. Endo, *Macromolecules* **1995**, *28*, 3042.

[5] J.-C. Lee, I. Tomita, T. Endo, *Polym. Bull.* **1997**, *39*, 415.

[6] J.-C. Lee, I. Tomita, T. Endo, *Macromolecules* **1998**, *31*, 5916.

[7] (a) J.-C. Lee, I. Tomita, T. Endo, *unpublished results*. (b) J.-C. Lee, I. Tomita, T. Endo, *Polym. Prepr. Jpn.* **1996**, *45*, 1463.

[8] (a) I. Tomita, J.-C. Lee, A. Nishio, T. Endo, *unpublished results*. (b) I. Tomita, A. Nishio, T. Endo, *Pacific Polym. Prepr.* **1993**, *3*, 661. (c) J.-C. Lee, I. Tomita, T. Endo, *Polym. Prepr. Jpn.* **1997**, *46*, 1617.

[9] (a) J.-C. Lee, I. Tomita, T. Endo, *unpublished results*. (b) J.-C. Lee, I. Tomita, T. Endo, *Polym. Prepr. Jpn.* **1998**, *47*, 1760.

[10] (a) J.-C. Lee, I. Tomita, T. Endo, *unpublished results*. (b) J.-C. Lee, I. Tomita, T. Endo, *Polym. Prepr. Jpn.* **1997**, *46*, 1619.

[11] I. Tomita, A. Nishio, T. Endo, *Appl. Organometal. Chem.* **1998**, *12*, 735.

[12] I. Tomita, A. Nishio, T. Endo, *Macromolecules* **1994**, *27*, 7009.

[13] Buntz et al. independently reported the synthesis of arylene-ethynylene type polymers from regioisomerically pure cyclobutadienecobalt derivatives. See: (a) M. Altmann, U. H. F. Bunz, *Macromol. Rapid. Commun.* **1994**, *15*, 785. (b) M. Altmann, U. H. F. Bunz, *Angew. Chem., Int. Ed. Engl.* **1995**, *34*, 569.

[14] (a) I. L. Rozhanskii, I. Tomita, T. Endo, *Macromolecules* **1997**, *30*, 1222. (b) I. L. Rozhanskii, I. Tomita, T. Endo, *Polymer* **1999**, *40*, 1581.

[15] (a) I. L. Rozhanskii, I. Tomita ,T. Endo, *Chem. Lett.* **1997**, 477. (b) Y. Sawada, I. Tomita, I. L. Rozhanskii, T. Endo, *J. Inorg. Organomet. Polym.* **2000**, *10*, 221. (c) Y. Sawada, I. Tomita, T. Endo, *Macromol. Chem. Phys.* **2000**, *201*, 510.

[16] Y. Sawada, I. Tomita, T. Endo, *Polym. Bull.* **1999**, *43*, 165.

[17] (a) Y. Sawada, I. Tomita, T. Endo, *unpublished results*. (b) Y. Sawada, I. Tomita, T. Endo, *Polym. Prepr. Jpn.* **1998**, *47*, 1445.

[18] I. Tomita, J.-C. Lee, T. Endo, *J. Organomet. Chem.* **2000**, *611*, 570.

[19] J.-C. Lee, I. Tomita, T. Endo, *Chem. Lett.* **1998**, 121.

[20] (a) J.-C. Lee, I. Tomita, T. Endo, *unpublished results*. (b) J.-C. Lee, I. Tomita, T. Endo, *Polym. Prepr. Jpn.* **1997**, *46*, 1617.

[21] (a) S. S. H. Mao, T. D. Tilley, *J. Am. Chem. Soc.* **1995**, *117*, 5365. (b) S. S. H. Mao, T. D. Tilley, *Macromolecules* **1997**, *30*, 5566.

[22] (a) K. Sato, Y. Nishihara, S. Huo, Z. Xi, T. Takahashi, *J. Organomet. Chem.* **2000**, *633*, 18. (b) S. Yamaguchi, R. Z. Jin, K. Tamao, F. Sato, *J. Org. Chem.* **1998**, *63*, 10060. (c) E. Block, M. Birringer, C. He, *Angew. Chem. Int. Ed. Engl.* **1999**, *38*, 1604. (d) D. Suzuki, R. Tanaka, H. Urabe, F. Sato, *J. Am. Chem. Soc.* **2002**, *124*, 3518. (e) R. Tanaka, S. Hirano, H. Urabe, F. Sato, *Org. Lett.* **2003**, *5*, 67.

[23] (a) M. Ueda, I. Tomita, *unpublished results*. (b) M. Ueda, I. Tomita, *Polym. Prepr. Jpn.* **2002**, *51*, 1259.

[24] (a) I. Tomita, K. Atami, T. Endo, M. Ueda, T. Utsumi, *unpublished results*. (b) I. Tomita, K. Atami, T. Endo, *Polym. Prepr. Jpn.* **1999**, *48*, 341. (c) I. Tomita, K. Atami, T. Endo, *Polym. Prepr. Jpn.* **2000**, *49*, 1641.

[25] (a) T. Utsumi, I. Tomita, *unpublished results*. (b) T. Utsumi, I. Tomita, *Polym. Prepr. Jpn.* **2002**, *51*, 267.

[26] (a) T. Utsumi, I. Tomita, *unpublished results*. (b) T. Utsumi, I. Tomita, *Polym. Prepr. Jpn.* **2002**, *51*, 1323.

[27] (a) M. Ueda, I. Tomita, *unpublished results*. (b) M. Ueda, I. Tomita, *Polym. Prepr. Jpn.* **2003**, *52*, 1255.

Macromol. Symp. **2004**, *209*, 231-251

Photochemically Degradable Polymers Containing Metal-Metal Bonds along Their Backbones: The Effect of Stress on the Rates of Photochemical Degradation

David R. Tyler,* Rui Chen

Department of Chemistry, University of Oregon, Eugene, OR 97403 USA
E-mail: dtyler@uoregon.edu

Summary: The syntheses of polymers that have metal-metal bonds along their backbones are described. The polymers are photodegradable because the metal-metal bonds homolyze when irradiated with visible light. The photochemical reactions of the polymers in solution are identical to the photochemical reactions of the discrete metal-metal bonded dimers. Typical reactions include metal-metal bond disproportionation and metal radical capture, for example, by chlorine atom abstraction from carbon tetrachloride. The polymers are also photochemically degradable in the solid state; thin films of the polymers degrade when irradiated with visible light in the presence of oxygen or in the absence of oxygen if the polymer backbone has a built-in radical trap. The origin of tensile stress-induced rate enhancements in the photodegradation of polymers was studied using the polymers with metal-metal bonds along their backbones and with built-in radical traps. By eliminating the need for external oxygen to act as a radical trap, the experimentally challenging problem of diffusion-controlled oxidation kinetics was avoided. Analysis of plots of quantum yields for degradation vs. stress reveals that stress increases the separation of the radical fragments produced by photolysis. An increased separation leads to less radical-radical recombination, which increases the efficiency of degradation. Quantitative knowledge of the factors that control polymer degradation rates will eventually allow synthesis of an ideal photodegradable polymer - one that has a tunable onset of degradation and that degrades quickly once degradation has started.

Keywords: metal-metal bonds; metal-polymer complexes; photochemical degradation; photochemistry; stress

Introduction

An interesting outcome of artificial weathering studies on polymers is the finding that tensile- and shear-stress can accelerate the rate of photodegradation.[1] Recent studies of this phenomenon have shown that stress will accelerate the photo-oxidative degradation of many

© 2004 WILEY-VCH Verlag GmbH & KGaA, Weinheim DOI: 10.1002/masy.200450517

polyolefins, including polystyrene,[2,3] polypropylene,[4-9] polyethylene,[10-13] polyethylene/ polypropylene copolymer,[14] and PMMA,[15] as well as polycarbonates,[8] nylon,[16] and acrylic-melamine coatings.[17] These findings are of enormous practical importance because most polymers are subjected to light and some form of temporary or permanent stress during their lifetime. It is thus important to find ways to control the enhanced degradation induced by the synergism of light and stress. In contrast, some polymers are designed to degrade when exposed to light. In order to control the onset of degradation and the rate of degradation in these materials, it is likewise important to understand the mechanistic origins of the synergism between light and stress in these systems. Our research is aimed at finding answers to two intriguing questions of scientific and practical concern: *why does stress cause changes in the photodegradation rate* and *what is the quantitative relationship between stress and the rate of photodegradation*? To answer these questions, we conceived and synthesized a new class of polymers containing metal-metal bonds along their backbones. This paper provides an overview of these new materials and it reports the results of our investigation on the effect of stress on photochemical degradation rates of polymers.

Background

A Workable Experimental Approach to the Problem

Very little is known about the quantitative relationship of stress to photochemical degradation rates. One of the reasons so little is known is that polymer degradations are mechanistically complicated.[18] This is not to say that the mechanisms are not understood; in fact, they are understood in detail.[18] Rather, the mechanisms are intricate, often involving multiple steps, cross-linking, and side-reactions; this makes pinpointing the effects of stress difficult. For example, one formidable complication is that oxygen diffusion is the rate-limiting step in many photooxidative degradations.[13],[19] This adds to the intricacy of the analysis because oxygen diffusion rates are frequently time-dependent.[19,20] To circumvent these experimental and mechanistic complexities and therefore make it less difficult to interpret data and obtain fundamental insights, we use three key experimental strategies in our study of stress effects on photodegradation. First, we study the problem using photodegradable polymers that contain metal-metal bonds along the backbone.[21-26] These polymers degrade with visible light by a straightforward mechanism involving metal-metal bond homolysis followed by capture of the metal radicals with an appropriate radical trap (typically either oxygen or an organic halide; Scheme 1). By studying the effect of stress on these "model" systems, we can extract information without the mechanistic complications inherent in the degradation mechanisms of organic radicals. (For example, metal radicals do not lead to crosslinking, so we can avoid this complicating feature found with organic radicals.) The second key experimental strategy is to use polymers that have built-in radical traps. By eliminating the need for external oxygen to act as a trap, we can exclude the complicating kinetic features of diffusion-controlled oxidation reactions. The third experimental strategy is to use the distinctive M-M bond chromophore to monitor spectroscopically the polymer photodegradations and to obtain quantum yields. Typically, degradation reactions are monitored by stress testing, molecular weight measurements, or attenuated total reflection (ATR) spectroscopy, all of which can be laborious and time consuming. Quantum yields on thin film polymer samples are easily measured in our lab with a computerized apparatus, and progress is expedited accordingly.

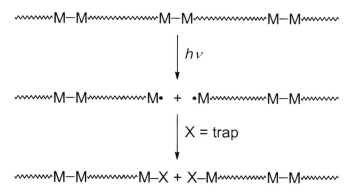

Scheme 1. Photochemical degradation of a polymer with metal-metal bonds along its backbone

Synthesis and Characterization of the Photodegradable Polymers

Our general synthetic route for incorporating metal-metal bonds into polymer backbones is based on the step polymerization techniques for incorporating ferrocene into polymer backbones.[27-33] Step polymers of ferrocene can be made by substituting the Cp rings with appropriate functional groups, followed by reaction with appropriate difunctional organic monomers (e.g., Equation 1).[34-36]

$$(1)$$

The analogous strategy for synthesizing metal-metal bond-containing polymers also uses difunctional, cyclopentadienyl-substituted metal dimers. A sample polymerization reaction is shown in Equation 2, which illustrates the reaction of a metal-metal bonded "diol" with hexamethylene diisocyanate (HMDI) to form a polyurethane.[21]

$$(2)$$

This step polymerization strategy is quite general, and a number of metal-metal bond-containing polymers have been made from monomers containing functionalized Cp ligands.[27-33, 37]

Synthesis of the Difunctional Dimers

The essential starting materials in the synthesis of the metal-metal bond-containing polymers are metal-metal bonded dimers with functional groups substituted on the Cp rings. The metal-metal bonds are relatively weak, and therefore they will not stand up to the harsh conditions typically required for the substitution of metal-coordinated Cp rings. For this reason, it is necessary to synthesize first the substituted cyclopentadienyl ligands and then coordinate these rings to the metals. A high yield synthetic route to the alcohol-substituted Cp rings is summarized in

Scheme 2. The syntheses of other substituted dimers are detailed in references.[21-23] A sample route to the difunctional dimer molecule is also shown in Scheme 2. Note that the general route shown for the Mo-containing dimers (involving Fe^{3+} oxidation of the anionic species) was developed by Manning.[38, 39]

Synthesis of the Polymers

Just as the comparatively weak metal-metal bonds pose problems for the synthesis of the difunctional dimers, they cause similar problems in the synthesis of the polymers. The relative weakness of the metal-metal bonds makes them more reactive than the bonds found in standard organic polymers; thus, under many standard polymerization reaction conditions, metal-metal bond cleavage would result. For example, metal-metal bonds react with acyl halides to form metal halide complexes. Therefore, the synthesis of polyamides using metal-metal bonded "diamines" and diacyl chlorides would simply lead to metal-metal bond cleavage rather than polymerization. Likewise, metal-metal bonded complexes are incompatible with many Lewis bases because the Lewis bases cleave the metal-metal bonds in disproportionation reactions.[40] This type of reactivity thus rules out many standard condensation polymerization reactions in which bases are used to neutralize any acids produced. Similar reasons prevent the use of acylchlorides in the synthesis of polyamides. All of the polymerization strategies are thus carefully designed to avoid cleaving the metal-metal bond during the polymerization process.

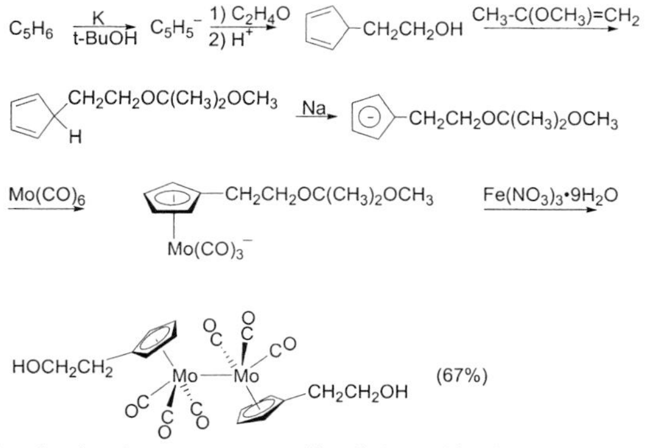

Scheme 2. Synthesis of an organometallic diol used in the polymer syntheses

A sample polymerization reaction, showing the synthesis of a polyurethane, was shown in Equation 3. Using similar synthetic strategies, various polyurethanes, polyureas (e.g., Equation 4), polyvinyls (e.g., Equation 5), and polyamides (e.g., Equation 6) were synthesized.[21-24] Note that the step polymers in the various equations have a metal-metal bond in every repeat unit. Experiments showed that it was not necessary to have a metal-metal bond in every repeat unit in order to photochemically degrade the polymers.[23] Copolymers are straightforwardly synthesized by adding appropriate difunctional organic molecules into the reaction mixture.

Another polymer synthesis strategy is to react the difunctional dimer molecules with prepolymers. Equation 3 shows an example of this technique.[23] (In this instance, the prepolymer is one of the Hypol polymers sold by W.R. Grace. Analysis of the sample showed it to contain, on average, three tolyl isocyanate end groups; M_n was about 2 000.)

$$(3)$$

Characterization of the Polymers.

The polymers were spectroscopically characterized by comparison of their infrared, electronic, and NMR spectra to model complexes.[21-24] For example, the product shown in Equation 8, a model complex for the polymer in Equation 3, was synthesized by reaction of $(CpCH_2CH_2OH)_2Mo_2(CO)_6$ with a *mono*isocyanate (Equation 4).

$$(4)$$

Typical M_n values, as measured by VPO or GPC, are between 5 000 and 20 000 (n = 7-25). Thus, in many cases, the polymers are best described as oligomers. However, it is important to note that no effort was made to maximize the molecular weights.

Photochemical Reactions in Solution

Irradiation of metal-metal bonded complexes into their lowest energy absorption band (≈ 500 nm) generally leads to one of three fundamental types of reactivity:[21, 41, 42] (1) The metal radicals produced by photolysis react with radical traps to form monomeric complexes (e.g., Equation 5). (2) The complexes react photochemically with ligands to form ionic disproportionation products (e.g., Equation 6). (3) The complexes react with oxygen to form metal oxides (Eqaution 7). (The latter reaction is likely a radical trapping reaction but may involve excited state electron transfer.) Higher energy excitation leads to M-CO bond dissociation. This type of reactivity is discussed below.

$$Cp_2Mo_2(CO)_6 + 2\ CCl_4 \xrightarrow{h\nu} 2\ CpMo(CO)_3Cl + 2\ [\cdot CCl_3] \qquad (5)$$

$$Cp_2Mo_2(CO)_6 + 2\ PR_3 \xrightarrow{h\nu} CpMo(CO)_3^- + CpMo(CO)_2(PR_3)_2^+ + CO \qquad (6)$$

$$Cp_2Mo_2(CO)_6 \xrightarrow[O_2]{h\nu} Mo\ oxides \qquad (7)$$

The qualitative photochemistry of the polymers in solution is analogous to the reactions of the discrete metal-metal bonded dimers in solution. [21-24] As in the photochemical reactions of the discrete dimers, the photochemical reactions of the polymers can be conveniently monitored by electronic absorption spectroscopy. The quantum yields for the reactions are in the range ≈ 0.1 to 0.6, depending on the specific polymer and the M-M bond.[22] Sample reactions of the polymers showing the three types of reactivity are shown in Equation 8-10.

$$\tag{8}$$

A fourth type of dimer reactivity is dissociation of a CO ligand from the dimer. Generally, this type of reactivity increases in efficiency relative to M-M photolysis as the radiation energy increases.[41] In solution, this type of reactivity generally leads to substitution. However, in the case of the $Cp_2Mo_2(CO)_6$ molecule, the reaction in Equation 11 occurs.[22] (Among the dimers, this reaction to form a triply bonded product is unique to the Mo and W species.)

$$\tag{9}$$

$$\left[-OCH_2CH_2 \cdots Mo-Mo \cdots CH_2CH_2O\overset{O}{\overset{||}{C}}NH(CH_2)_6NH\overset{O}{\overset{||}{C}}-\right]_n \xrightarrow[O_2]{h\nu} \text{metal oxides} \qquad (10)$$

$$Mo-Mo \;\underset{}{\overset{UV}{\rightleftharpoons}}\; (CO)_2Mo\equiv Mo(CO)_2 + 2\,CO \qquad (11)$$

An analogous photoreaction occurs with polymers containing the Mo-Mo unit (Equation 12).

$$\left[-OCH_2CH_2 \cdots Mo-Mo \cdots CH_2CH_2O\overset{O}{\overset{||}{C}}NH(CH_2)_6NH\overset{O}{\overset{||}{C}}-\right]_n$$

$$\updownarrow \text{UV, }-2\,CO$$

$$\left[-OCH_2CH_2 \cdots (CO)_2Mo\equiv Mo(CO)_2 \cdots CH_2CH_2O\overset{O}{\overset{||}{C}}NH(CH_2)_6NH\overset{O}{\overset{||}{C}}-\right]_n \qquad (12)$$

In both reactions 11 and 12, addition of CO to the product solution causes the system to back-react to reform the starting materials. Once again, the main point to be made is that the solution photochemistry of the polymers is analogous to the solution photochemistry of the discrete metal-metal bonded dimers.

Photochemical reactivity in the absence of exogenous radical traps is possible in the case of polymers that have carbon-halogen bonds along their backbones. For example, irradiation of polymers I – III in solution in the absence of CCl_4 or O_2 led to net metal-metal bond cleavage.[43] Spectroscopic monitoring of the reaction showed that metal-metal bond cleavage is accompanied

240

by an increase in the concentration of CpMo(CO)$_3$Cl units. Photochemical reactions analogous to that in Scheme 3 were proposed.

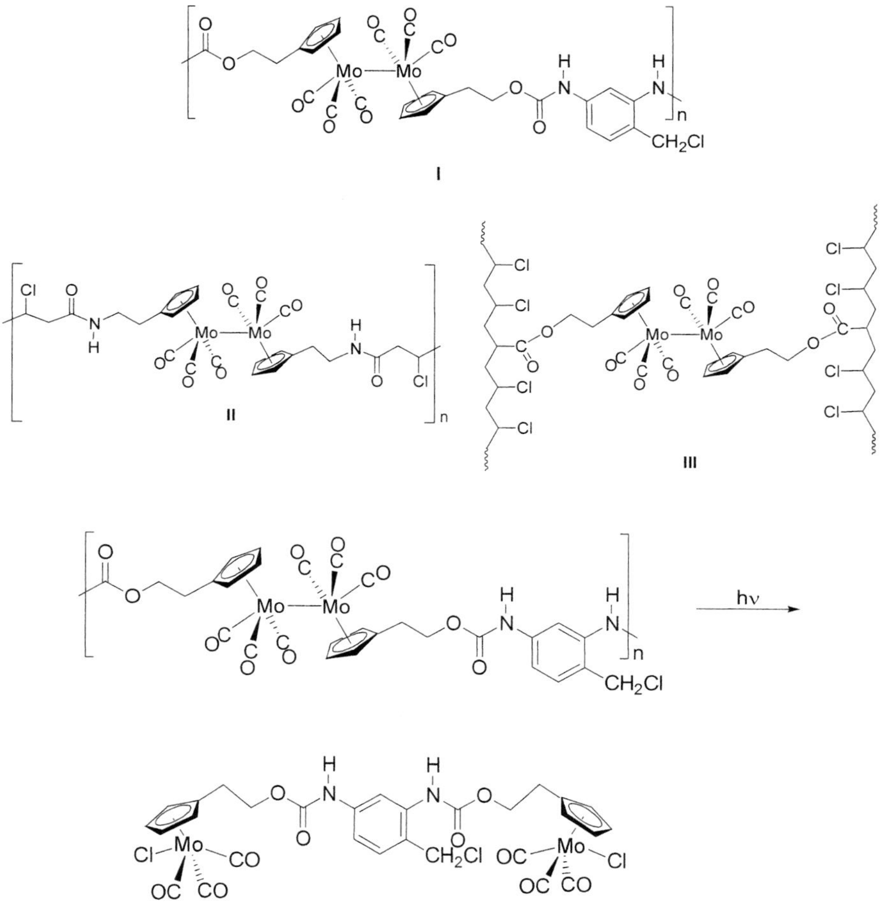

Scheme 3. Photochemical reaction of polymer III in the absence of an external trapping reagent

Photochemistry in the Solid State

Thin films of the polymers ($\approx$ 0.05 mm in thickness) containing metal-metal bonds along their backbones reacted when they were exposed to visible light, whether from the overhead fluorescent lights in the laboratory, from sunlight, or from the filtered output of a high pressure Hg arc lamp. [21-24] All of the films were irradiated both in the presence and absence of oxygen. For each film and its dark reaction control, the absorbance of the d$\pi\rightarrow\sigma^*$ transition near 500 nm was monitored periodically over a period of several months. Only the polymer films that was exposed to sunlight in air completely degraded, while thin films stored in the dark in air or irradiated under nitrogen showed only slight degradation over a one-year period. From these results, it was concluded that the decomposition of the polymers requires both light and air (oxygen). Infrared spectra of the decomposition products showed the absence of products with CO ligands, as indicated by the absence of any stretches in the region 1 600-2 200 cm^{-1}. As mentioned previously, oxide complexes form in the solution phase reactions of $Cp_2Mo_2(CO)_6$ with O_2, and it was proposed that the metal-containing decomposition product of the polymer is a metal oxide.

These data suggest that oxygen is necessary for the solid-state photochemical reaction to occur. It was proposed that oxygen traps the metal radicals produced in the photolysis of the metal-metal bonds, thereby preventing radical recombination (Equation 13). If oxygen diffusion is rate-limiting then the relative rates of oligomer photochemical decomposition in the solid-state would reflect the oxygen diffusion rate.

$$\text{~~~Mo–Mo~~~} \xrightarrow{\;h\nu\;} \text{~~~Mo• •Mo~~~} \xrightarrow{\;O_2\;} \text{~~~MoO}_n \quad \text{O}_n\text{Mo~~~} \qquad (13)$$

As described in the previous section, polymers I - III were designed to degrade in the absence of exogenous radical traps by building-in carbon-chlorine bonds along their backbones. As indicated, all of these polymers did degrade in the absence of oxygen when dissolved in solution. Likewise, all three polymers degraded in the solid state when irradiated in the absence of oxygen.[43]

Theories of Stress-dependent Photodegradation.

Mechanistic hypotheses to explain stress-accelerated photodegradation fall into three main categories. These categories are illustrated by reference to Scheme 4.

$$\text{wwM-Mww} \underset{k_{recombination}}{\overset{\phi_{homolysis}}{\rightleftharpoons}} \text{wwM}\bullet \ \bullet\text{Mww} \xrightarrow{X = trap} 2 \ \text{wwM-X}$$

Scheme 4. A generalized reaction scheme showing photolysis of a bond along the backbone in a polymer (M represents a generic atom, carbon or otherwise)

In one category, it is proposed that stress changes the quantum yields of the reactions that lead to bond photolysis, i.e., it is proposed that $\phi_{homolysis}$ varies with stress. The second category attributes the variation in degradation rates with stress to changes in the efficiency of radical recombination following homolysis, i.e., $k_{recombination}$ is proposed to depend on stress. And last, the third category attributes the effects of stress to changes in the rate of the radical trapping reaction. The salient points of these hypotheses are outlined in the following sections.

Stress-induced Changes in $\phi_{homolysis}$.[44]

For a direct photochemical bond cleavage, the photochemical step in Scheme 4 labeled "$\phi_{homolysis}$" can be broken down into the nominal set of elementary steps shown in Equation 14. (The asterisk in Equation 14 is used to indicate an excited state of the molecule.)

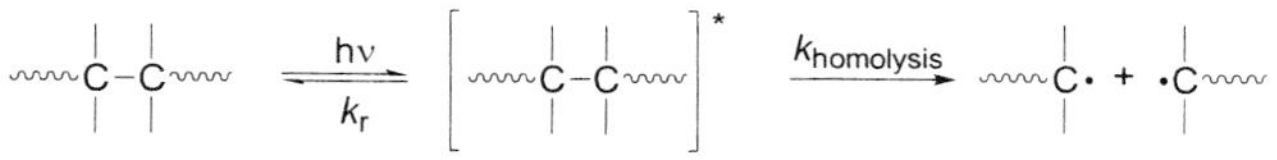

$$\text{www}\overset{|}{\underset{|}{C}}-\overset{|}{\underset{|}{C}}\text{www} \underset{k_r}{\overset{h\nu}{\rightleftharpoons}} \left[\text{www}\overset{|}{\underset{|}{C}}-\overset{|}{\underset{|}{C}}\text{www} \right]^* \xrightarrow{k_{homolysis}} \text{www}\overset{|}{\underset{|}{C}}\bullet \ + \ \bullet\overset{|}{\underset{|}{C}}\text{www} \tag{14}$$

Equation 14 shows the value of $\phi_{homolysis}$ (in Scheme 4) in terms of the rate constants in Equation 15. Clearly, if stress affects either k_r or $k_{homolysis}$ then $\phi_{homolysis}$ will vary with stress.

$$\phi_{homolysis} = \frac{k_{homolysis}}{k_{homolysis} + k_r} \tag{15}$$

No studies have investigated the effect of stress on k_r, but Plotnikov derived a theory for the stress dependence of $k_{homolysis}$.[44] His quantitative hypothesis attributes the increase in degradation rates with applied stress to a decrease in the activation barrier for bond dissociation in the excited state. The decrease in the activation barrier is shown pictorially in the energy state diagrams in Figure 1, which compare an unstressed bond to a stressed bond for the case of an adiabatic photochemical reaction.[45]

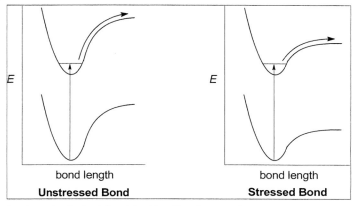

Figure 1. The photophysical origin of the Plotnikov hypothesis

Analysis of these energy surfaces led to the following equation:[44]

$$k_{homolysis} = k_0 \exp(-E_a/T) \qquad \text{where } E_a = D\left[(1 - \kappa)^{1/2} - \frac{\kappa}{2} \ln \frac{1 + (1 - \kappa)^{1/2}}{1 - (1 - \kappa)^{1/2}}\right] \qquad (16)$$

and where $\kappa = f/F_m$, f is the stretching force on the bond, $F_m = \alpha D^*/2$, $\alpha = \omega[\mu/2D]^{1/2}$, μ is the reduced mass, ω is the bond vibration frequency, and D and D^* are the bond dissociation energies in the ground and excited states, respectively. Similar equations were also derived for the case of predissociative- and nonadiabatic-mechanisms. For each type of photochemical reaction, the theory predicts that an increase in stress will increase the quantum yield of degradation but will eventually level off, i.e., further increases in stress will not increase the quantum yield (Figure 2). The Plotnikov hypothesis has not been tested experimentally.

244

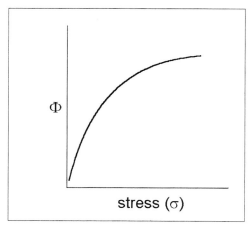

Figure 2. A plot of quantum yield for degradation vs. stress according to the Plotnikov equation

Stress-induced Changes in $k_{recombination}$

A number of authors[10, 11, 46-48] have proposed theories for explaining the stress dependence of polymer photochemical degradation rates that are based on the concept of stress affecting the ability of geminate radical pairs, formed in bond cleavage reactions, to recombine. These various theories differ slightly in their details, but they are similar overall and are discussed together here. For convenience in referring to these theories, the general concept is given the name "decreased radical recombination efficiency" (DRRE) hypothesis.

In the DRRE hypothesis, the effect of stress on the photochemical reactions of polymers is divided into four stages (Figure 3). Stage one is the low stress domain. In this stage, there is little or only slight deformation of the original polymer structure and the rate of photodegradation is not greatly affected (Figure 3).

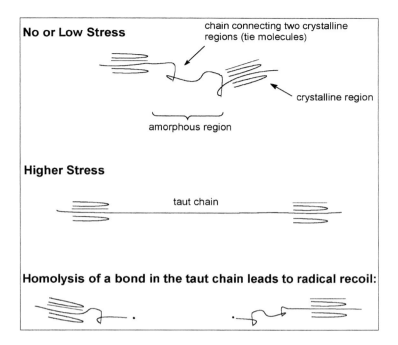

No or Low Stress

chain connecting two crystalline regions (tie molecules)

crystalline region

amorphous region

Higher Stress

taut chain

Homolysis of a bond in the taut chain leads to radical recoil:

Figure 3. The proposed effect of stress in the several stages of the "decreased radical recombination" hypothesis

In stage two, higher stress causes significant morphological changes, including the straightening of the polymer chains in the amorphous regions. These straightened chains contain taut tie molecules. (Tie molecules are the interlamellar- or intercrystal-fibrils.) When bonds in the taut tie molecules are cleaved by light, the probability of radical recombination is decreased relative to non-stressed samples because entropic relaxation of the chain drives the radicals apart and prevents their efficient recombination because of their increased separation. At slightly higher stresses (stage 3), the chains are not only straightened but "stretched," and mechanical recoil also aids in the separation of the radicals (much like the mid-points of a stretched spring would fly apart if it were cut in the middle). According to this model, the role of stress is to increase the separation of the radical fragments produced by photolysis. An increased separation leads to slower radical-radical recombination, which increases the probability of radical trapping and thus of degradation.

Finally, in stage four (not shown in Figure 3), a strong stress is present, which gives the polymer a fibrillar structure with a higher degree of orientation and crystallinity. (In this stage, ordered regions develop as segments of different chains align.) Diffusion in a crystalline structure is retarded relative to the amorphous material, and the efficiency of degradation is expected to decrease because of decreased diffusion apart of the radical pair and decreased radical-trap mobility. In summary, the DRRE theory predicts that tensile stress will initially increase the quantum yield of degradation and then further increases in stress will decrease the quantum yield (Figure 4).

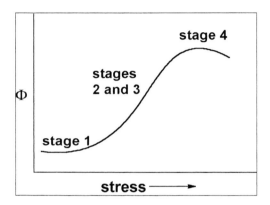

Figure 4. A plot of quantum yield for degradation vs. stress according to the decreased radical recombination hypothesis

A note of caution when interpreting data is that, if one experimentally observes a decrease in the quantum yield as predicted for stage four behavior, it is important to establish that the origin of the decrease is in fact a higher degree of orientation and crystallinity. An alternative explanation is that microcracks and fissures have formed in the sample and these are acting to relieve the stress, which in turn would also decrease the quantum yields. Nguyen and Rogers detected this alternative mechanism in their study of acrylic-melamine coatings.[49]

Stress-induced Changes in the Rates of Radical Reactions Subsequent to Radical Formation - the Zhurkov Equation

The effect of stress on reaction rates that occur subsequent to the formation of the radical caged pair is generally summarized by noting the Zhurkov equation. This empirical equation is based on Zhurkov's observation that the rates of thermal degradations show an exponential dependence on the stress.

$$rate = A \ exp[-(\Delta G - B\sigma)/RT] \tag{17}$$

In this equation, ΔG is an "apparent" activation energy, σ is the stress, and A and B are constants. This equation is empirical but it has been found to describe the behavior of many thermal degradation reactions. Because the Zhurkov equation is similar in form to the Arrhenius equation,[50] this gave credence to the suggestion that stress alters the (effective) activation energy of the degradation reactions. (Recall that the Arrhenius equation shows the relationship between the activation energy for an elementary step and the rate constant for that elementary step.) The Zhurkov equation (for thermal reactions) does not specifically fit into any of the previous three categories because it deals with an "effective" activation energy, which is a composite of the activation barriers for the $\phi_{homolysis}$, $k_{recombination}$, and $k_{trapping}$ steps.

The suggestion has been made that a Zhurkov-like equation might also apply to photochemical degradation reactions.[1] A version of the equation suitably modified for a photochemical reaction has never appeared explicitly in the literature, but the following general equation is implied:

$$\Phi_{obs} = A \ exp[-(\Delta G - B\sigma)/RT] \tag{18}$$

(Several authors note that the photochemical equation should contain the light intensity as a variable.[1] In the equation above, the intensity dependence of the *rate* is found in the quantum yield, which recall is the rate divided by the intensity. Additional intensity dependences might show up in the A and B terms.) A plot of quantum yield as a function of stress for a system that follows Zhurkovian behavior is shown in Figure 5.

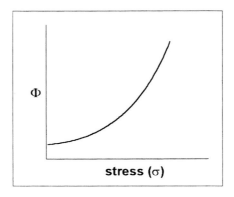

Figure 5. A plot of quantum yield for degradation vs. stress according to the Zhurkov equation

Experimental Studies of Stress on Photochemical Degradation Efficiencies

Thin films of polymer **III** were photochemically reactive ($\lambda > 500$ nm) in the absence of oxygen.[43] A Cl-atom abstraction reaction analogous to the reaction in Scheme 3 was proposed because infrared spectroscopic monitoring of the photochemical reaction showed the disappearance of the $\nu(C\equiv O)$ bands of the $Cp_2Mo_2(CO)_6$ moiety at 2 009, 1 952, and 1 913 cm^{-1} and the appearance of bands attributed to the $CpMo(CO)_3Cl$ unit at 1967 and 2047 cm^{-1}. The application of tensile stress changed the photodegradation efficiency, and a plot of relative quantum yield vs. stress is shown in Figure 6. Note that tensile stress initially caused the quantum yield to increase, but after a certain point additional stress caused a decrease in the quantum yield.

These results are consistent with the "decreased radical recombination" (DRE) hypothesis. All three hypotheses discussed in the previous section predict that stress will initially increase the efficiency of degradation, but only the "decreased radical recombination" hypothesis predicts that further increases in stress will eventually cause a decrease in photochemical efficiency. The conclusion, at least in this one system, is that the role of stress is to increase the separation of the photochemically generated radical pair, which decreases their probability of recombination. To the authors, knowledge, this is the first and only experimental confirmation of the behavior predicted by the "decreased radical recombination" hypothesis in which oxygen diffusion is not a complicating factor. (Note that x-ray scattering and infrared spectroscopy experimentally confirmed the increase in chain order in the stressed PVC polymer used in these experiments.)

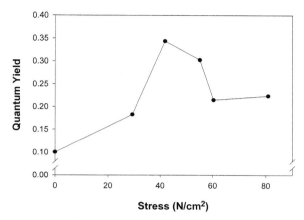

Figure 6. Quantum yields for degradation of **III** vs. applied tensile stress

Further evidence for the role of the DRRE mechanism in determining how stress affects degradation rates comes from the effect of plasticizer on a polymer. Because plasticizers increase chain-end mobility, one would predict that a plasticized polymer should have a higher quantum yield for degradation, at any given stress, compared to the unplasticized polymer at the same stress. Likewise, because chain alignment should be facilitated, the maximum in the curve of Φ vs. stress should occur at a lower stress. To test these predictions and to further test the DRRE hypothesis on which they are based, the plasticizer DOP was added to polymer II (25% plasticizer) and the effect of stress on Φ was measured. (The new material had a glass transition temperature of less than 25 °C.) As predicted, the quantum yields at any particular stress were higher for the plasticized material and the maximum in the curve occurred at lower stress.[51] Both results are consistent with the DRRE hypothesis.

Summary

Step polymers containing metal-metal bonds along their backbones can be synthesized by reacting difunctional, cyclopentadienyl-substituted metal carbonyl dimers with appropriate difunctional organic molecules. A synthesis of a polyurethane was shown, but a wide variety of polymers, including polyureas and polyamides, have been synthesized by analogous routes using appropriate starting materials. Although not discussed, it is noteworthy that chain polymers containing metal-metal bonds can be synthesized from metal-metal dimers containing Cp ligands with vinyl group substituents. The polymers are photodegradable because the metal-metal bonds homolyze when irradiated with visible light. The photochemical reactions of the polymers in solution are identical to the photochemical reactions of the discrete metal-metal bonded dimers. Typical reactions include metal-metal bond disproportionation and chlorine atom abstraction from carbon tetrachloride. The polymers are also photochemically degradable in the solid state; thin films of the polymers degrade when irradiated with visible light in the presence of oxygen or if the polymer backbone has a built-in radical trap. Tensile stress increases the rate of polymer photodegradation. The interpretation of the plots of quantum yield for degradation vs. stress is that the role of stress is to increase the separation of the radical fragments produced by photolysis. An increased separation leads to less radical-radical recombination, which increases the efficiency of degradation. Quantitative knowledge of these and other factors that control polymer degradation rates will eventually allow synthesis of an ideal photodegradable polymer - one that has a tunable onset of degradation and that degrades quickly once degradation has started.

Acknowledgments

Acknowledgment is made to the Donors of the Petroleum Research Fund, administered by the American Chemical Society, and to NSF (DMR-0096606) for the support of the author's work reported herein.

[1]   J. R. White, N. Y. Rapoport, *Trends in Polym. Sci. (Cambridge, U.K.)* **1994**, *2*, 197.
[2]   B. O'Donnell, J. R. White, *J. Mat. Sci.* **1994**, *29*, 3955.
[3]   B. O'Donnell, J. R. White, *Polym. Prepr. (Am. Chem. Soc., Div. Polym. Chem.)* **1993**, *34*, 137.
[4]   L. Tong, J. R. White, *Polym. Degrad. Stab.* **1996**, *53*, 381.
[5]   R. Baumhardt-Neto, M. A. De Paoli, *Polym. Degrad. Stab.* **1993**, *40*, 59.
[6]   R. Baumhardt-Neto, M. A. De Paoli, *Polym. Degrad. Stab.* **1993**, *40*, 53.
[7]   G. E. Schoolenberg, P. Vink, *Polymer* **1991**, *32*, 432.

[8] C. T. Kelly, L. Tong, J. R. White, *J. Mat. Sci.* **1997**, *32*, 851.
[9] B. O'Donnell, J. R. White, *Polym. Degrad. Stab.* **1994**, *44*, 211.
[10] W. K. Busfield, M. J. Monteiro, *Mater. Forum* **1990**, *14*, 218.
[11] D. Benachour, C. E. Rogers, *ACS Symposium Series* **1981**, *151*, 263.
[12] M. Igarashi, K. L. DeVries, *Polymer* **1983**, *24*, 1035.
[13] A. Huvet, J. Philippe, J. Verdu, *Eur. Polym. J.* **1978**, *14*, 709.
[14] W. K. Busfield, P. Taba, *Polym. Degrad. Stab.* **1996**, *51*, 185.
[15] F. Thominette, J. Verdu, *Polym. Prepr. (Am. Chem. Soc., Div. Polym. Chem.)* **1994**, *35*, 971.
[16] M. Igarashi, K. L. DeVries, *Polymer* **1983**, *24*, 769.
[17] H. Nguyen Truc Lam, C. E. Rogers, *Polym. Mater. Sci. Eng.* **1987**, *56*, 589.
[18] G. Geuskens, *Compr. Chem. Kinet.* **1975**, *14*, 333.
[19] A. V. Cunliffe, A. Davis, *Polym. Degrad. Stab.* **1982**, *4*, 17.
[20] J. Malik, A. Hrivik, D. Q. Tuan, *Advances in Chemistry Series* **1996**, *249*, 455.
[21] S. C. Tenhaeff, D. R. Tyler, *Organometallics* **1991**, *10*, 473.
[22] S. C. Tenhaeff, D. R. Tyler, *Organometallics* **1991**, *10*, 1116.
[23] S. C. Tenhaeff, D. R. Tyler, *Organometallics* **1992**, *11*, 1466.
[24] G. F. Nieckarz, D. R. Tyler, *Inorg. Chim. Acta* **1996**, *242*, 303.
[25] G. F. Nieckarz, J. J. Litty, D. R. Tyler, *J. Organomet. Chem.* **1998**, *554*, 19.
[26] D. R. Tyler, J. J. Wolcott, G. F. Nieckarz, S. C. Tenhaeff, *ACS Symp. Ser.* **1994**, *572*, 481.
[27] C. U. Pittman, Jr., M. D. Rausch, *Pure Appl. Chem.* **1986**, *58*, 617.
[28] K. Gonsalves, L. Zhan-Ru, M. D. Rausch, *J. Am. Chem. Soc.* **1984**, *106*, 3862.
[29] K. E. Gonsalves, R. W. Lenz, M. D. Rausch, *Appl. Organomet. Chem.* **1987**, *1*, 81.
[30] F. W. Knobloch, W. H. Rauscher, *J. Polym. Sci.* **1961**, *54*, 651.
[31] C. U. Pittman, Jr., *J. Polym. Sci., Polym. Chem. Ed.* **1968**, *6*, 1687.
[32] K. E. Gonsalves, M. D. Rausch, *J. Polym. Sci., Part A: Polym. Chem.* **1988**, *26*, 2769.
[33] W. J. Patterson, S. P. McManus, C. U. Pittman, Jr., *J. Polym. Sci., Polym. Chem. Ed.* **1974**, *12*, 837.
[34] P. Nguyen, P. Gomez-Elipe, I. Manners, *Chem. Rev. (Washington, D. C., U.S.)* **1999**, *99*, 1515.
[35] I. Manners, *Adv.Organomet. Chem.* **1995**, *37*, 131.
[36] I. Manners, *Coord. Chem. Rev.* **1994**, *137*, 109.
[37] M. Moran, M. C. Pascual, I. Cuadrado, J. Losada, *Organometallics* **1993**, *12*, 811.
[38] R. Birdwhistell, P. Hackett, A. R. Manning, *J Organomet. Chem.* **1978**, *157*, 239.
[39] The difunctional metal complex dimers were characterized by the usual spectroscopic methods, and it is worth noting that the electronic absorption and infrared spectra in the $\nu(C\equiv O)$ region are virtually identical to those of the unsubstituted dimers. The (CpCH$_2$CH$_2$OH)$_2$Mo$_2$(CO)$_6$ dimer was further structurally characterized by X-ray crystallography. See S. C. Tenhaeff, D. R. Tyler, and T. J. R. Weakley, *Acta Crystallogr.* **1991**, *C47*, 303.
[40] A. E. Stiegman, D. R. Tyler, *Coord. Chem. Rev.* **1985**, *63*, 217.
[41] T. J. Meyer, J. V. Caspar, *Chem. Rev. (Washington, D.C., U. S.)* **1985**, *85*, 187.
[42] G. L. Geoffroy, M. S. Wrighton, *Organometallic Photochemistry*, **1979**.
[43] R. Chen, M. Yoon, R. Tyler David, *unpublished work.*
[44] V. G. Plotnikov, *Doklady Akademii Nauk SSSR* **1988**, *301*, 376.
[45] See ref. [44] for the appropriate pictures for non-adiabatic reactions and for dissociative excited states.
[46] T. L. Nguyen, C. E. Rogers, *Polym. Mater. Sci. Eng.* **1985**, *53*, 292.
[47] E. Baimuratov, D. S. Saidov, I. Y. Kalontarov, *Polym. Degrad. Stab.* **1993**, *39*, 35.
[48] Y. A. Shlyapikov, S. G. Kiryushkin, A. P. Marin, *"Antioxidative Stabilization of Polymers"*, Taylor and Francis, Bristol, PA **1996**.
[49] T. L. H. Nguyen, C. E. Rogers, *Polymer Science and Technology (Plenum)* **1988**, *37*, 431.
[50] T. H. Lowry, K. S. Richardson, *"Mechanism and Theory in Organic Chemistry". 3rd Ed*, **1987**.
[51] R. Chen, D. R. Tyler, manuscript in preparation.

RETURN TO: **CHEMISTRY LIBRARY**
100 Hildebrand Hall • 510-642-3753

LOAN PERIOD	2 *2-HR USE*	3
4	5	6

ALL BOOKS MAY BE RECALLED AFTER 7 DAYS.
Renewals may be requested by phone or, using GLADIS,
type **inv** followed by your patron ID number.

DUE AS STAMPED BELOW.

FORM NO. DD 10
3M   5-04

UNIVERSITY OF CALIFORNIA, BERKELEY
Berkeley, California 94720–6000